T0287835

Praise for *TRANSFORMED*

"*TRANSFORMED* should be in the hands of every organizational leader evolving their organization in this post-pandemic world. Marty has captured all the impactful aspects of product operating model and I can wholeheartedly attest, based on my own personal journey in both retail and healthcare, this is a key reference guide moving forward."

—*Prat Vemana, Chief Digital and Product Officer, Target*

"Marty has been helping set the standard in Product for years with *INSPIRED, EMPOWERED*, and his work at SVPG. With *TRANSFORMED*, Marty offers a powerful guide to help companies master the product concepts, overcome the inevitable obstacles of a transformation, and learn from real-world examples of product transformations. This book should be a trusted resource for any company trying to transition to the product model."

—*Tyler Tuite, Chief Product Officer, Carmax*

"SVPG understands what most executives don't. In order to achieve organizational transformation, you must experience a personal transformation. This book will show you how to do both."

—*Brendan Wovchko, Chief Technology Officer, Ramsey Solutions*

"If you have—or are seeking—the courage to transition to a product operating model, *TRANSFORMED* will be your guiding light. This book not only underscores why your organization should transform but also provides practical techniques to implement the change and drive positive results. Marty's candid and straightforward advice on pressing questions, such as how to overcome objections from within the organization, is indispensable and actionable. If you've ever questioned your company's ability to innovate—or stay relevant—in an era of exponential technological growth, *TRANSFORMED* is a must-read."

—*Melissa Cohen, Senior Product Director, Rocket Mortgage*

"If you feel like your organization needs to evolve to the product operating model, then *TRANSFORMED* is a must-read. In typical Marty fashion, he candidly outlines why to do it, how to do it, the challenges you'll face (both the *who* and the *what*), and how to navigate and overcome them. It's a quick and powerful read and one you'll come to revisit over and over."

—*Shawn Boyer, CEO, GoHappyLabs*

"No one has as much depth and breadth of experience with world-class product teams as Marty. In *TRANSFORMED*, Marty presents a very pragmatic path for otherwise-successful companies that still haven't cracked the nut on consistently building high-impact software products. This path is not just about adopting one magic tactic for goal setting, or rolling out a popular software development process. Instead, *TRANSFORMED* presents a more holistic approach for organizational design, strategy, culture, leadership, and ongoing improvement that actually works in the real world."

—*Shreyas Doshi, Advisor and former product leader at Stripe, Twitter, Google, Yahoo*

"SVPG has already taught us how the product model empowers all companies to better serve their customers. *TRANSFORMED* now teaches us how to unlock that power in any company despite its industry, product type or past management approach. With this book, we now all have the roadmap needed to transform our businesses into product powerhouses."

—*Michael Newton, CEO, Qorium; former VP Product at Nike*

"*TRANSFORMED* is the definitive resource for executives looking to gain the ultimate competitive edge in transforming from a feature factory to a world-class product organization that delivers unrivaled value to customers."

—*Michelle Longmire, CEO, Medable*

"In an era where technology is the heartbeat of most successful businesses, *TRANSFORMED* stands out as the essential guide for any company navigating the transition to the product operating model. Marty masterfully demystifies the journey, blending hard-won insights with compelling case studies, making it not only a roadmap for transformation but also an inspiration. Having witnessed countless organizations grapple with this very shift, I can confidently say this book is the beacon they've been waiting for. A must-read for anyone serious about harnessing the power of technology to propel their business forward."

—*Mike Fisher, former CTO at Etsy*

"If you're looking to understand the importance of a product-first approach to business, this book is a must-read. *TRANSFORMED* is an inspirational guide through the product model that speaks directly to the non-product executive."

—*Matt Brown, Chief Financial Officer, Altair Engineering, Inc.*

"Marty Cagan and the SVPG team have done it again. Through the use of great examples, and their signature style of hitting issues head-on, they've created a guide to help companies navigate the roadblocks and pitfalls that come with transforming to the product operating model. This book should be required reading for any executive team who wants to make their journey more successful."

—*Anish Bhimani, Chief Product Officer, Commercial Banking, JPMorgan Chase*

"The money, time, and patience you can save by implementing the insights in this book is incredible. This was the missing piece in Marty's collection that now addresses the common mistakes and patterns you encounter while transforming from the rest to the best. I hope every board, CEO, and product team reads this."

—*Anuar Chapur, VP of Technology, The Palace Company*

"*TRANSFORMED* is the field guide to embedding innovation into organizational DNA. With real-world lessons, provided in easily understood narratives, Marty and team bring decades of experience to navigating product transformation. A must-read for any CEO or CIO who is looking to build a smart, responsive, and technologically enabled organization."

—*Madhu Narasimhan, EVP Innovation, Wells Fargo*

"There are so many ingredients necessary to make a successful transformation work. Marty gives you everything you need in *TRANSFORMED*. From a valuable perspective for the CEO and Board level, to critical technology components like CI/CD, to how product can effectively partner across the organization and beyond, this is the path to success."

—*Thomas Fredell, Chief Product Officer, ShiftKey*

"Building great products that customers love but work for your business is *really, really* hard. Transforming an organization to discover and deliver impactful products consistently is *even harder*. Meeting Marty in 2019 and reading his book *Inspired* started my journey as a product leader, transforming an organization from an uninspired feature factory to an empowered organization. Christian Idiodi delivering a transformation workshop to our entire organization in 2020 accelerated that process and led us to exponential growth, thriving through the pandemic and then coming out even stronger after the pandemic. *TRANSFORMED* powerfully and simply articulates the principles and components that were critical to successfully transforming our business to the product model and then scaling it."

—*Ronnie Varghese, Chief Digital Officer, Almosafer*

"This is a must-read for any company leader interested in learning what it takes to successfully implement and sustain the product operating model. *TRANSFORMED* not only provides a taxonomy that decodes how the best technology companies work, but also supplies the reader with specific techniques to drive a true digital transformation and showcases real examples from several companies across diverse industries. I have no doubt that this book will become an essential change management guide to drive product transformations across the world!"

—*Juan D. Lopez, Director of Product Management, Blue Origin*

"Every CEO and Chief Product Officer who wants to delight customers, increase revenues and profits, and attract and retain top talent should read and apply this book—it will transform everything."

—*Phyl Terry, Founder and CEO, Collaborative Gain*

"*TRANSFORMED* is the guide for a successful digital transformation. It's comprehensive and practical, giving companies the blueprint of what great looks like and how to get there. The book is a must-read not only for businesses in the midst of transformation, but for any company aspiring to operate at its best."

—*Gabrielle Bufrem, Product Coach*

"Companies spend huge amounts of time and money trying to transform, and all too often get little in return. *TRANSFORMED* shows what's possible on the other side, what it *actually* takes to pull it off, and what can derail the effort. I finally have a book

I can give to the people I work with that clearly and practically explains the journey we're on."

—*Andrew Skotzko, Product Leadership Coach & Advisor*

"If your organization aims to innovate like the best tech companies, you need to read *TRANSFORMED*. I'm buying copies for all of my clients."

—*Felipe Castro, Founder, OutcomeEdge*

"Ever wondered what sets companies like Stripe, Slack, and Apple apart? It's the Product Operating Model, a dynamic conceptual framework that transcends traditional processes. In *TRANS-FORMED*, discover how you, as an executive or product leader, can embrace the model's set of first principles, reflecting what the best product companies believe to be true about creating products. Recognize that there's no one-size-fits-all approach to product development. Here, strategic direction is woven into the fabric of your organization, with product leaders in tech, design, and product providing essential strategic context and dedicating themselves to coaching and mentoring their staff. Learn to accelerate value delivery and to empower teams to tackle problems to solve, rather than features to build. Marty, with 20 years of transformation expertise, offers a practical guide to instill trust, innovation, and adaptability. Your journey taking your organization to the next level begins now."

—*Marcus Castenfors, Partner at Crisp & Co-author*
of Holistic Product Discovery

TRANSFORMED

The Silicon Valley Product Group Series

INSPIRED: How to Create Tech Products Customers Love, 2nd Edition
(Marty Cagan, 2017)

EMPOWERED: Ordinary People, Extraordinary Products
(Marty Cagan with Chris Jones, 2021)

LOVED: How To Rethink Marketing For Tech Products
(Martina Lauchengco, 2022)

TRANSFORMED: Moving To The Product Operating Model
(Marty Cagan with Lea Hickman, Christian Idiodi, Chris Jones, and
Jon Moore, 2024)

HOW TO TRANSFORM YOUR COMPANY TO
INNOVATE LIKE THE WORLD'S TOP COMPANIES

MARTY CAGAN

WITH **LEA HICKMAN, CHRISTIAN IDIODI, CHRIS JONES,** AND **JON MOORE**
Silicon Valley Product Group

TRANSFORMED

MOVING TO THE
PRODUCT OPERATING
MODEL

WILEY

For general information on our other products and services or for technical support, please contact our Customer Care Department within the United States at (800) 762–2974, outside the United States at (317) 572–3993 or fax (317) 572–4002.

Wiley also publishes its books in a variety of electronic formats. Some content that appears in print may not be available in electronic formats. For more information about Wiley products, visit our web site at www.wiley.com.

Library of Congress Cataloging-in-Publication Data:

Names: Cagan, Marty, author.
Title: Transformed : moving to the product operating model / Marty Cagan,
    with Christian Idiodi, Chris Jones, Lea Hickman, and Jon Moore.
Description: Hoboken, New Jersey : John Wiley & Sons, Inc., [2024]
Identifiers: LCCN 2023049041 (print) | LCCN 2023049042 (ebook) | ISBN
    9781119697336 (hardback) | ISBN 9781119697404 (adobe pdf) | ISBN
    9781119697398 (epub)
Subjects: LCSH: Organizational change. | Product management. | New
    products.
Classification: LCC HD58.8 .C24 2024 (print) | LCC HD58.8 (ebook) | DDC
    658.4/06—dc23/eng/20231019
LC record available at https://lccn.loc.gov/2023049041
LC ebook record available at https://lccn.loc.gov/2023049042

COVER DESIGN: PAUL MCCARTHY

SKY10062829_011824

This book is dedicated to Bruce Williams (1950–2016). Bruce was the person who gave me the encouragement and confidence to leave my comfort zone of leading product for other companies and to go out on my own and start the Silicon Valley Product Group (SVPG).

In addition to being an exceptional human who truly helped improve the lives of countless people, he was the best friend that anyone could ever hope for. Bruce provided me everything from office space to expert advice on design, sales, marketing, publishing, finance, and more.

Without Bruce, it's likely I would have stayed a product leader and not had the opportunity to write any books or to enjoy the previously unimaginable quality of life that I've experienced as part of SVPG. I thank Bruce for encouraging me to make the single best career decision I have ever made.

I wish that everyone could be lucky enough to have a Bruce Williams in their lives.

Contents

PART

I

Introduction

This book has three *big* goals:

First, we want to *educate* you as to exactly what the product operating model is and what it means to work in this way.

Second, we want to *convince* you—primarily with detailed case studies of successful transformations—that, while such transformation is difficult, it is absolutely possible for you to transform your company to the product model.

Third, we want to *inspire* you—primarily with several truly impressive case studies of product innovation—of what you too will be capable of once you successfully transform your organization.

We start by discussing why we wrote *TRANSFORMED* and how it relates to the books that led the way to this one: *INSPIRED* and *EMPOWERED*.

Next, we lay out the main reasons that companies are motivated to embark on this very significant amount of time and effort to transform.

Then we jump in and describe an all-too-typical transformation effort, which usually yields the same all-too-predictable—and disappointing—results.

It's very important that we provide you with a realistic understanding of what is involved and required for a successful transformation. Many people will try to tell you it's easy—if you simply hire them to help. The reality is very different.

But it's also important that we convince you that it is indeed very possible to succeed. And further, we want you to understand what you'll be able to do as a company once you have developed these new muscles.

Finally, we tackle the dilemma of knowing you need to change, but not necessarily having leaders who have worked this way before.

CHAPTER

1

Who Is This Book For?

TRANSFORMED is written for anyone who is trying to help their company move to the product operating model.

That includes:

- Members of product teams who are trying to move to the product model
- Product leaders trying to guide their company through these changes
- Senior executives of the company, especially the CEO and CFO, who want to learn what is involved in moving to the product model, to decide if this is something they wish to pursue, and if so, how they can help
- Other company executives, stakeholders, and employees who are impacted by the product model, and want to learn how to constructively engage
- Product coaches who are trying to help companies successfully transform to the product model

One very common question is: "We're not a tech company, so is the product model even relevant for us?"

This is a very widespread confusion. When our industry refers to a "tech company," the term is *not* referring to what the company *sells*, it is referring to how the company believes it should *power its business*.

Tesla sells cars. Netflix sells entertainment. Google sells advertising. Airbnb sells vacation lodging. Amazon started with selling books and now sells just about everything.

The difference is not *what* these companies sell; it's *how* they design and build what they sell, and how they run their business.

The product operating model is for companies that believe they should be powering their business with technology. Today, that's a very wide range of businesses across virtually every industry.

How does this book relate to *INSPIRED* and *EMPOWERED*?

We wrote *INSPIRED* to share the best practices and techniques used by the top product teams operating in the product model.

Then we wrote *EMPOWERED* to share the best practices and techniques used by the top product leaders to provide their product teams with the environment they need to thrive in the product model.

Yet, by far the most common question we get is: "Our company operates so differently from what you describe. Is it even possible for a company like ours to change to the product model, and how would we change?"

The goal of this book is to answer that question.

For the past 20 years, we at SVPG have been helping companies through these changes.

What we've learned is that there are some companies—mostly newer companies that were created in the internet era—born out of this new model of technology-powered product companies. For them, this way of working is quite natural.

However, for the vast majority of companies in the world, the ways of working we describe are very foreign.

One CEO described the experience of transforming as "changing from driving on the right side of the road to the left—but gradually."

These companies know they need to transform to compete in an era of rapidly changing enabling technology, but they have found that the degree of change necessary is substantial, and anything but easy.

If you've read *INSPIRED* or *EMPOWERED*, then you'll recognize the components of the product model, but those books don't address how to make the necessary changes to how you work.

If you have not read our other books, this book was written to provide the necessary understanding of the concepts of the product model and give you the foundation you need.

2

What Is a Product Operating Model?

Unfortunately, the tech industry is not great when it comes to standardization of terms. We often have different terms for the same concept and, in some cases, different definitions of the same term.

We don't like introducing new nomenclature, but we do need to be clear on what we mean when we talk about important concepts in this book.

The most fundamental concept we use and need to be clear on is the general *product operating model*.

We did not invent the term—we've found it in use in several strong product companies, and several more use the shortened form *product model*.

The product operating model is not a process, or even a single way of working. It is a *conceptual model* based on a set of first principles that strong product companies believe to be true. Principles such as the necessary role of experimentation, or prioritizing innovation over predictability.

We're using the term in this book to refer to the way of operating that we find the best product companies consistently use.

Essentially, the product operating model is about consistently creating technology-powered solutions that your customers love, yet work for your business.

From the financial perspective, it's about getting the most out of your technology investment.

It's important to emphasize that there is **no single right way to build products**. There will be more on this critical point later. But for now, know that there are many good ways, much as there are an even larger number of bad ways.

It's also a good time to point out something we emphasize in each of our books: **Nothing you will read in this book was invented by us.** We simply write about what we observe in the best product companies. That said, we don't write about everything we see; we focus on what we believe are the common themes and first principles. We are effectively playing the roles of *curator* and *evangelist*.

The product operating model is sometimes referred to as *product-led company* or *product-centric company*. We don't like those terms because they often have the unfortunate side effect of implying that the product organization is taking over.

Similarly, while the product operating model is an example of a *customer-driven company*, that term has been misused to the point that it has lost its utility.

A tougher concept to name is the model you might be coming *from*.

The main influence on how the company thinks of its previous model is generally this: Who is really in the driver's seat?

In many companies, you'll hear constant references to "the business" in terms of making requests, and they're running what is commonly referred to as the *IT model* (as in "IT is there to serve the business").

A close cousin of the IT model is the *project model*, where the CFO plays an outsize role because funding and staffing are typically project based.

If general stakeholders are in the driver's seat, then it's often called the *feature-team model* (because each stakeholder drives their roadmap of features).

If sales is driving, then you'll often hear the term *sales-driven product*, and if marketing is driving, you'll hear *marketing-driven product*.

Throughout this book, we refer to whatever model you are coming *from* as your *prior model*, and we refer to the model you're working to transform *to* as the *product operating model*, or simply the shortened term *product model*.

In part III of this book, **Transformation Defined,** you'll see a detailed definition of what it means to work in the product operating model.

There are several other important product terms and concepts that we discuss in this book, and we will define those as we go.

What Is a Product?

This seemingly simple question comes up very frequently, but there are several layers to this question, and there are different reasons that people ask the question.

Sometimes the person is concerned they are not working on something that is customer-facing, such as an e-commerce app or a consumer device. Maybe they are working on an internal tool, used by employees at their company to take care of the company's paying customers. Or maybe they're working on a platform service that is used to build other types of products. Or maybe a back-office–style system that provides critical data. Or maybe the person is building a device and not purely software. We have different terms for these types of products, but rest assured, these are all products that the product model is designed to help with.

Sometimes the question is whether what you're working on constitutes a *complete* product, or if it's just a small piece of something much larger. This is getting at the important topic of something we call *team topology*, but for our purposes, these are all considered products as well, even if just a small part of a larger whole.

And sometimes people are asking the question because they are not actually building anything. Plenty of types of products are not technology-powered products built by engineers. Perhaps the person manages a relationship with a third-party vendor, and they are wondering if the product model applies to them as well. In this case, no. The product model is meant to address the unique challenges of those who are *building* technology-powered products and services.

CHAPTER

3

Why Transform?

By making the decision to read this book, odds are you already have your reasons for embarking on a transformation.

But with the stakes so high, and the effort required so substantial, it's helpful to have a clear articulation of your purpose.

Generally speaking, we find companies are motivated by one or more of three major drivers:

A Competitive Threat

Few industries remaining have not already been attacked by new competitors providing demonstrably better solutions for customers.

As of this writing, the latest disruptive technology, Generative AI, is already reshaping the competitive landscape for many industries. As is always the case, some companies are leveraging this new technology to solve long-standing problems for their customers in ways that are just now possible, and other companies are being left behind.

This process of disruptive innovation has taken place in industries as varied as financial services, health care, retail, automotive, logistics, advertising, and even space exploration.

Certain industries have higher switching costs, more significant regulatory hurdles, or a degree of government protection, so this disruption has taken longer, but few companies feel truly secure.

These companies realize they are competing in a battle for customers with weapons and strategies that once worked but that today cannot match the tools and capabilities of their new adversaries.

A Compelling Prize

Other companies are motivated by the financial rewards available to those companies that have demonstrated the ability to consistently innovate on behalf of their customers.

They see the valuations of the new generation of companies, as well as the rewards to those that have successfully transformed, and the promise of this prize motivates investors, executives, product leaders, and employees to pursue this path.

Frustrated Leaders

Some companies pursue the product model because they have grown frustrated at the level of funding they've been pouring into their technology efforts, with so little in return.

The constant cost overruns, the disappointing results, the time it takes to ship anything, customers who are losing patience, and the endless stream of blaming and excuses.

Perhaps the leaders have read about companies that spend less yet generate much more, and they wish to learn if these methods might apply to them.

In some cases, one or more additional factors conspire to push a company to the point where its leaders feel they must take the leap to transform:

- Perhaps you find your most capable technology employees leaving your company out of frustration with the current way of operating.

- Perhaps your company has hired a leader who comes from a strong product company and has been encouraging others to consider this approach.

- Perhaps your customers' expectations have changed, and while they wish to remain loyal, they are increasingly vocal about their frustrations with your offerings and your pace of improvement.

Whatever the motivations, fundamentally, companies transform because *they believe they need to be able to successfully take advantage of new opportunities that arise and effectively respond to serious threats.*

CHAPTER

4

A Typical Transformation

Most companies that we've engaged with on a transformation have already tried to transform—at least once—before they realized they've spent a lot of time and money going nowhere.

The story that follows is all too common. We've removed the names, but the circumstances are very real. See how much of their story you recognize.

This company had dominated the financial operations market since its inception in 2000. It had always had a strong customer focus, priding itself on being able to deliver exactly what the customer requested, which ended up driving a long series of "specials" to meet the perceived needs of different individual customers.

Initially, this focus on doing whatever was necessary to close the deal achieved some real business success. Internally and externally, delivering quickly on client needs became a key brand strength.

That is, until it wasn't.

Despite increasing headcount and a culture of long working hours, the time to deliver new capabilities had noticeably slowed. At the same time, there were several new entrants to the market. Each had moved fast to gain traction. The result was that sales had started to lag, and market share was dropping.

In response, the company pursued a strategy of aggressive acquisition, moving quickly to purchase and integrate additional solutions. But far from igniting a new era of profitability, the acquisitions proved problematic and were not generating the hoped-for revenue.

A New Start

The board reacted, taking action to make changes at the very top.

The company's incoming CEO arrived with an impressive commercial track record. After quickly making her[1] way through the ranks of a global consultancy to become its youngest senior partner, she then assumed a variety of C-level roles across numerous commercial finance organizations.

As a first-time CEO, she understood that her career would be defined by her results. This was no time to fail, and she had done her homework. The company still had strong market share, a recognized brand name, and was profitable. But she knew change was required— she had pitched as much to the board. Her appointment was based on a mandate to reinvigorate the brand.

The playbook was well rehearsed, one she had practiced across multiple different sectors in her years as a management consultant. She quickly got to work diagnosing the problems.

The company had become a sprawling organization, with offices across the globe, including in the US, India, and Europe. The resulting cultural differences ran deep.

After years of acquisition-led growth, the company was essentially several companies, with little integration, heavy technical debt, and frustrated customers.

---

[1]Aligned with our long-standing interest in expanding the perception of who a strong leader is, we use she/her pronouns throughout this book. That being said, this book is intended for *anyone* of *any* gender identity. Our view is that if you are a good person who cares about the craft of product, we hope to have you as a friend.

The new CEO's previous relationships allowed her to quickly employ the advice of consultancy services, which helped to rapidly diagnose some key legacy practices.

The changes would need to be significant. They included a new, centralized organizational structure, the insourcing of engineering, a more focused business mission, and the creation of a new product unit housing product managers and designers. A new chief digital officer (CDO) was hired to lead the transformation effort.

Friction and Pushback

Numerous challenges were taken on in parallel, not least the insourcing of engineering. Hiring a team of 600 engineers in tough market conditions meant high salaries, and 12 months in, the company had exceeded the budget but was still short of its required numbers. To keep costs down, rather than hire skilled product managers, the company opted to retitle a team of business analysts. The rationalization was that the product suite was complex and required deep domain experience.

Change was proving to be more disruptive and costly than the company had originally anticipated. Sales, long regarded as the sole owner of customer success, was not happy with the changes and complained directly to the CEO. While the new CDO reported positively on new engineering hires and the creation of multiple product teams, things were moving slowly, and costs continued to increase.

Away from the ears of the executives, the incoming engineers gasped at the levels of technical debt. Each one of the acquired businesses had continued to operate on its legacy technical stack. There had been no serious attempt to integrate their solutions. Many of the original people were gone. Those who were left could barely keep many of those systems running. The engineers saw no real alternative to immediate replatforming.

Privately, leadership was sympathetic to the new engineers, but there was little appetite to take on yet another significant effort, this time for replatforming. They were encouraged to refactor where they could but to try to find a way to accelerate the roadmap delivery.

After two years of transformation, things were moving slower than ever. The CEO decided that the core issue was with engineering, and they needed help to deliver predictable and reliable results.

Focus on Predictability

The CIO recommended a switch to a new, more formal and structured engineering delivery process, and teams were retrained on this process in hopes of increasing predictability, albeit with longer, less frequent release cycles.

But with even fewer releases, the new battleground was which business unit could shoehorn in sufficient amounts of features to satisfy the pain—and cost—of customer upgrades. There was also significant new process overhead.

With all eyes on each and every lengthy release, a team of program managers was hired to oversee the delivery process by adding a layer of governance and control.

As costs continued to increase, releases became few and far between. Meanwhile, competitors were thriving. The board's initial expectations of success were now all but forgotten.

Additionally, a key number of the recently hired engineers had resigned, citing a failure of leadership. Customer retention was proving increasingly difficult, with many frustrated by the all-too-obvious lack of progress.

Believing they had nothing to lose, a small number of product and engineering leaders chose to speak directly to the CEO.

Their technology, they argued, was preventing the company from meeting the needs of customers, and the level of bureaucracy was now so high that they felt no ability to make any forward progress.

None of the options was good. The technology mismanagement of the past had finally caught up with the company.

With tens of millions of dollars in transformational costs spent, the CEO wondered what the company had to show for it. After four years of pain and expense, and with the combined frustration of her fellow executives bearing down on her, the CEO was informed she no longer had the support of the board.

The company's transformation had, gradually, then all at once, failed.

CHAPTER
5

The Role of the CEO

Probably the most important lesson we have learned in 20 years of helping companies with transformations to the product model is the absolutely critical role that the CEO needs to play in order for the transformation to succeed.

Please don't misunderstand. The CEO does not need to have prior experience with the product model. And the CEO doesn't have to spend a lot of time on the transformation itself.

But, nevertheless, the CEO role is critical.

Now, every CEO will of course say she supports the transformation.

Yet most don't truly realize what this means until much later in the effort.

Here is the problem: Transformation impacts the company far beyond what was previously your technology department.

The issue is that transformation impacts sales, marketing, finance, HR, legal, business development, compliance, and manufacturing.

In this book, we discuss precisely how and why transformation impacts each of these areas. But the important point here is that not every key leader and stakeholder will be enthusiastic about moving to this new way of working.

Most will be motivated to at least try out an approach that holds the potential for better results. And many of these people realize that learning to work effectively in the product model is increasingly an asset to any resume.

It's also important to realize that even the most enthusiastic stakeholder or executive may have questions or very legitimate objections that they want to make sure are addressed.

But some will passively resist, and likely a few will actively resist. Either because they are protecting their current responsibilities, or because of the age-old preference for the devil you know versus the devil you don't.

These problematic stakeholders all eventually report up to the CEO, and that's whom they will look to decide if this is truly important and necessary.

As the legendary coach Bill Campbell used to say: "The company cares about what the leader cares about."

Too often, the CEO delegates ownership of the transformation to a CIO, a chief digital officer, or a chief transformation officer. And while these individuals can influence decisions and actions *within* their organization, unless the stakeholders report to them (which is rarely the case), this same problem will result.

If you're a CEO, this may be the first time you're learning that transformation is something that goes beyond IT. If that's the case, it's critical that you consider this new information.

It's true that most companies begin their transformation efforts by focusing on the necessary changes to product, design, and engineering. And that's appropriate because until and unless those capabilities are established, the rest is premature.

There are new competencies you'll need to develop, new skills and new principles to instill in your organization, and, most important, a cultural change that needs to take place.

However, too often it is only after those changes are made that the company realizes that taking advantage of these new capabilities impacts the *entire* company. And when the CEO and the other senior leaders are not actively supportive of these changes, the transformation effort stalls out.

To be explicit: *The CEO needs to be viewed as the chief evangelist for the product model.*

If this is something your CEO is unwilling or unable to do, then you can likely save yourself a lot of time, money, and effort by reassessing your readiness.

However, the good news is that a company operating successfully under the product model improves the situation for everyone, not just the product and engineering organization.

Employees feel more pride in the company's products. Marketing has more to promote and position. Sales can sell more. Everyone sees the financial impact. Employee morale and retention improve.

Which is all to say that there's a strong case to be made to justify the active help and support of the CEO.

CHAPTER

6

A Guide to *TRANSFORMED*

We hope that by now you are starting to develop an appreciation for what we mean when we say that transformation is hard.

This book is intended to prepare you for what lies ahead.

One thing you *won't* find in this book is any sort of recipe or playbook for transformation. Many people out there will try to sell you such a thing, but unfortunately, we have never seen these one-size-fits-all, overly simplified approaches work.

Interspersed throughout the book are **Transformation Stories**—case studies of successful transformations—as well as **Innovation Stories**—case studies of innovative solutions created by companies that have previously transformed.

The transformation case studies are meant to give you confidence that, although difficult, *success is absolutely possible*. These stories are shared with you by the person who was leading the product organization at each of these companies.

The innovation case studies are intended to get you excited about what you'll be able to do once you've transformed.

Beyond the case studies, the book is organized into the following parts:

We start things off in part II, **Transformation Defined**, with a comprehensive definition of what is truly meant by transformation to the product operating model.

The next two parts get to the heart of the new skills and capabilities you'll need to develop to move to the product model.

In part III, we begin by discussing the new **Product Model Competencies** you will need to develop in order to transform. If you think you already have these competencies in your company, you are almost certainly wrong and headed for transformation failure. Don't be fooled by people who have simply adopted these titles.

Next, in part IV, we present the **Product Model Concepts**—and the product model principles they are based on—that are the foundation of the product model. Most companies very quickly realize that they don't yet have these skills, which is the first step to learning them.

The new product model competencies and the product model concepts are the foundation upon which transformation rests.

In part VI, **The Product Model in Action**, we discuss how the product organization constructively and effectively partners with customers, sales, product marketing, finance, stakeholders, and the senior executives.

Next, in part VIII, we discuss the **Transformation Techniques** that are helpful in guiding your organization through a change of this magnitude. Change is always difficult, but there are several key techniques and tactics designed to facilitate transformation. We begin with an organization assessment, then we describe a variety of tactics. We also describe the importance of ongoing transformation evangelism.

You may already realize there's a bit of a catch-22 going on here. How does a company that has never worked in a new way, and may not have any leaders who have worked that way, learn this new way of working?

Books and various types of training can help (if the authors and trainers know what they're talking about), but they are rarely sufficient. We explore how you can move to the new model when your leaders have never worked this way before.

Next, in part X, **Overcoming Objections**, we discuss the various concerns that commonly arise from each key stakeholder—sales, marketing, customer success, finance, HR, the CIO, the PMO, the CEO, and the board—and also the objections that come from inside the product organization.

These are all legitimate objections and concerns, and people typically raise them in good faith, simply because they can see an issue and they don't yet know how that issue will be addressed. We consider each of these objections and then discuss how to overcome them.

In part XI, **Conclusion**, we try to connect the dots and summarize the critical points discussed in the book, including the common themes from the transformation case studies and the keys to success.

Tough Love for Product Leaders

Successful transformation is hard to pull off, and it requires tackling tough issues where often the people involved really don't want to hear it.

So, in that spirit, we'll start with a bit of that tough love right now, intended for the product leaders:[1]

There is no question that the executives of the company have a major role to play in helping the company transform.

However, the surprise for so many product leaders is that they have *at least* as much work to do.

It is truly perplexing how many product leaders believe that once the senior executives change their ways, all will be good. They focus their energy on getting others to change, rather than on changing themselves.

(continued)

---

[1]Just to be very clear on our use of terms: Product managers, product designers, tech leads, and the other engineers, data scientists, and user researchers are all *individual contributors*. The term "product leaders" refers to the *managers* of product management, product design, and engineering. The term "executives" refers to the company's *senior leaders*—the CEO, CFO, COO, CMO, CRO, and so on.

They complain about having weak product managers.

They complain about not being empowered.

They complain about engineers who don't seem engaged.

They complain about having stakeholders that don't trust them.

They complain about CEOs who demand detailed product roadmaps.

Yet, they don't seem to realize that each of these problems is very much a consequence of their own actions (or inactions, as the case may be).

If a product leader is not willing to take responsibility for raising the skill level of her product people, or to correct hiring mistakes, then the product teams will be staffed with people who lack the necessary level of competence.

When that's the case, who would be surprised if the teams are not empowered?

Who would be surprised if the engineers act like mercenaries?

Who would be surprised if the stakeholders don't trust the product managers?

Who would be surprised if the CEO doesn't trust the people responsible for those weak product people?

More generally, there are two important things for you as product leader to keep in mind as you embark on a transformation: First, you have as much ownership as you have credibility; and second, part of your job as a product leader is to change hearts and minds.

While it is true that the executives have much to change to move to the product model, even larger changes are required of the product leaders and product teams.

Our intention is to make very clear the changes that are needed from *all* sides, but success starts with the product organization raising its game.

PART

Transformation Defined

There are so many anti-patterns when it comes to transformation.

Many of us have witnessed failed transformations, but few have witnessed true successes. Which makes the lessons learned from successful transformations unusually valuable.

One question we get quite frequently is: "What can our company do after transformation that we couldn't do before?"

Since we in the product world talk a great deal about outcomes and results, we think that's the right question.

And this book emphasizes the capabilities and results of those companies that have transformed, especially their ability to respond to threats and take advantage of new opportunities.

But assuming the leaders of a company believe they need to truly transform, *what does that really mean?*

Many of you have heard things like "Transforming to Agile may be necessary, but it's in no way sufficient."

Or, "The heart of transformation is moving from feature teams to empowered product teams."

Or, "The goal is to transform to a product-led company."

While each of these comments may speak to a specific aspect of transformation, none gives a good, holistic picture of what is truly meant by *transformation*.

So, in this book, we take a different approach.

Instead of applying a label ("Agile," "Empowered Teams," "Product-Led Company"), we think it's more useful to look at what is actually changing.

In this book, we discuss the product model as change along three distinct dimensions:

1. Changing how you build
2. Changing how you solve problems
3. Changing how you decide which problems to solve

Changing How You Build

Despite Agile being around for so many years, far too many companies still are stuck doing monthly or quarterly (or worse) big-bang releases.

The rise of so-called *fake Agile*[1] lets these companies fool themselves into thinking they're working the way they need to without actually improving *how* they build in any meaningful way.

Our company and our customers need us to provide a reliable service that they can depend on.

This means frequent, small releases. It means instrumenting your technology so that you know that it's working and how it's being used. It means monitoring your technology so that you can, hopefully, detect any problems before your customers do. And it means being able to prove that new capabilities deliver the necessary value before you deploy widely.

---

[1] The most egregious example of what is known as "fake Agile" is SAFe, but even teams using basic Scrum yet releasing only monthly or quarterly are missing out on the actual benefits of Agile.

If you're not doing continuous delivery, then at the very least you need to be doing true releases[2] no less frequently than every two weeks.

Changing How You Solve Problems

When a company talks about transforming from feature teams to empowered product teams, they mainly mean changing how they solve problems.

Instead of stakeholders prioritizing their perceived solutions (features and projects) and providing these in the form of a roadmap to a feature team, now the product team is assigned problems to solve, and the product team is empowered to discover a solution that is valuable, usable, feasible, and viable.

The point is to push the decisions on the best solution down to the people closest to the enabling technology and to the users interacting with this technology.

In practice, this means developing the skills to rapidly test product ideas to discover a solution worth building (this is called *product discovery*) and ensuring that your engineers and product designers have a true product manager (skilled in the customers, the data, the business, and the industry) so that the product team has the necessary cross-functional skills to succeed.

It's important to note that this change implies a new relationship with the company stakeholders—where the product team moves from a subservient relationship to a collaborative relationship, and where the product team must discover a solution that customers love yet works for the business.

---

[2]The meaning of a *true release* varies in different types of products, but for our purposes, it means getting to the state where you can successfully deliver new capabilities into the hands of your customers.

Changing How You Decide Which Problems to Solve

Getting skilled at product discovery so that your teams can consistently and quickly solve hard problems in ways that your customers love, yet work for your business, is a significant leap forward by anyone's definition. But it doesn't address the question: "How did you decide this was the most important problem to be solved?"

In prior models, stakeholders usually decide which problems get solved. In the product model, a new and critical competency is the job of product leadership.

Every company faces several threats, and many opportunities. Which threats you take seriously, and which opportunities you decide to pursue, can mean the difference between success and failure.

A strong product company has a compelling product vision and an insight-based product strategy to identify the most critical problems that need to be solved to deliver on the business objectives.

A couple of important notes:

First, all three of these dimensions—changing how you build, how you solve problems, and how you decide on which problems to solve—depend on strong product leaders (leaders of product management, product design, and engineering), and we hope you can start to get a sense of why product leadership is so hard. The people on the product teams will require coaching and strategic context.

Second, at a high level, you can treat each of these three dimensions as steps in a progression, and indeed this would be one way to approach a company transformation. But, in reality, each of these three dimensions is a spectrum, and you can and should make progress on different dimensions in parallel. Much more on this in part VIII, **Transformation Techniques**.

To summarize:

- Changing how you build and deploy means moving from big, quarterly releases to a cadence of small, consistent, frequent releases. This is key to consistently improving time to market.

- Changing how you solve problems means moving from stakeholder-driven roadmaps and feature teams to empowered product teams given problems to solve, and then using product discovery to come up with solutions that are valuable, usable, feasible, and viable. Doing this means solving both for the customer and for your business. This change is key to consistently improving time to money.

- Changing how you decide which problems to solve is typically the most profound change of all, as it drives what opportunities you choose to pursue and how you make the most out of your investment, including product vision and product strategy. This change is key to maximizing the return on your technology investment.

Hopefully this framing can help you think more holistically about the transformation that your company may need to make, and where you may be on that journey.

In the next three chapters, we drill into each of these three dimensions and describe *why* these changes are so important to strong product companies.

While these concepts may be somewhat technical, a deep understanding of technology is not required to understand the importance—and the value—to your business and your customers.

7

Changing How You Build

Your technology investment is all about creating value for your customers and your company.

There are several important aspects to creating this value, but at the end of the day, the primary skill you depend on to build your products is your engineers.

For many technology-powered companies, the engineers are the single largest line-item cost.

In most of the prior models, typically each technology effort is treated as a project.

Each project is funded, staffed, planned, executed, and delivered. The project ends upon delivery, and people roll off to other work.

Realize that everything we build has two outputs that could create value: what we make and what we learn. In the project model, we lose most of what we learn.

If and when we want to work on that area again, we spend time and money relearning things we already paid for once. Or, more likely, not learning them, and making costly mistakes.

Further, teams that have to live in the code they build treat it differently than teams that know they are rolling off at the end of the project. This is why technical debt is so rampant in the project model.

It's similar to the difference between remodeling a house to sell it versus to live in it. In the former case, just painting over the wallpaper is faster and cheaper, and who cares if the paint peels later?

It's not a coincidence that this is the way outsourcing is done, and we often say if this is how you want to work, then you may as well hire Accenture because that company is better at working this way than you are ever likely to be.

But in the product model, this way of working is far too expensive in terms of time and money, and, more important, it almost never delivers the innovation your customers and your company need.

So, this is why transformation starts with changing *how* you build. And this means changing the focus from *projects* to *products*.

In the product model, products are instead managed as ongoing efforts—improving every week (in strong product companies, many times a day), typically for several years.

You may change the work of the teams from adding new capabilities to generating additional revenue, or reducing operating costs, or doing whatever the company needs. But, in general, the team continues to invest in the products until either the company decides to stop investing and just sustain the current revenue, or the decision is made to sunset the product from its offering altogether.

In this model of continuous development, you need small, frequent, reliable release vehicles.

Strong product companies learned many years ago that while this may sound counterintuitive, the data very clearly show that the more you build, and the more changes you are delivering, it is much better for you—and especially for your customers—to release more frequently rather than less frequently.[1]

If you genuinely care about providing a reliable service for your customers, it is much easier to ensure that a small number of changes

---

[1] See *Accelerate: The Science of Lean Software and DevOps: Building and Scaling High Performing Technology Organizations* by Nicole Forsgren, Jez Humble, and Gene Kim (IT Revolution Press, 2018).

are working properly and don't introduce any inadvertent problems than it is to group a large number of changes together and try to deliver all at once (referred to in the industry as a *big-bang* release).

Just to make this important point very clear, if each of your product teams is not releasing at least once every two weeks, then you will not be able to take care of your customers as you need to.[2]

Moreover, for everything you deploy, you need to ensure that the new capabilities are instrumented so that you know the product is operating properly and so you understand how your customers are actually using your product. You also need to be continuously monitoring your product to detect any issues, hopefully before your customers do. For many important changes, you need to be able to prove that the new capability is providing the necessary value before you deploy broadly (the standard way of doing this is with an A/B test).

You may think that this doesn't apply to your type of product, or you may point out that your customers are pushing you for *less frequent* releases, not more frequent releases, but in chapter 18, "Product Delivery," we discuss the reasons why this is critical for both you and your customers, the principles that drive this way of working, and the mechanisms used to do this.

A Note About Agile and Agile Coaches

You have almost certainly heard of Agile. A main reason so many companies moved to Agile processes over the past 20 years is that these methods were intended to provide the forcing function to get product teams to achieve these consistent, small, frequent releases.

It's true that moving to small, frequent releases can require a significant investment—mainly in testing and deployment automation. But for those companies that employed Agile methods to get them to

(continued)

---

[2]You will hear many common excuses about why your company can't do this, and we address them in part X, **Overcoming Objections.**

release *no less frequently than every other week*, it has provided them and their customers real value.

That said, it's important to understand that you don't need Agile methods to have consistent, small, frequent releases.

In fact, many of the best product teams in the world have mastered the ability to consistently deploy very small releases (referred to as *continuous integration and continuous deployment, or CI/CD*), yet they don't follow any formal Agile process or methods.

Contrast this with organizations that invest so much time and money in adopting Agile coaches, rituals, roles, methods, and processes yet, at the end of that considerable expense and time, are still releasing quarterly and inflicting the associated pain on their customers.

And just to be perfectly clear on this very critical point: If your company is still releasing yearly, quarterly, or even monthly, it doesn't matter how many Agile rituals you follow or how many so-called Agile coaches you may employ. The truth is that you are *not* Agile (or even lowercase agile) in any meaningful sense, you are not getting the benefits, and you will not be able to serve your customers or your business as you need to.

Many companies spent many millions of dollars moving to Agile processes because they thought this is what it meant to transform, yet they have few if any meaningful results to show for the effort. If this is you, we're sure you are beyond frustrated.

But the truth is that whether or not you choose to fly an Agile flag, you need to get each of your product teams to the point where they can deliver frequent, small, reliable releases no less than once every two weeks.

If your current people can't do that, you will need to bring in either experienced engineering leaders or engineers or a true product delivery coach to show them how.

CHAPTER

8

Changing How You Solve Problems

Changing how you build, test, and deploy is important no matter what you choose to build, but too many companies just become more efficient feature factories.

They ship more new features than ever, yet they don't see the corresponding value to their customers, or the impact to their business.

In fact, in most companies today, the percentage of features and projects on product roadmaps that actually generate a positive return on investment is depressingly small. Most industry analysts place it in the range of 10 to 30 percent.

If you compare the list of capabilities that your company needs with the impact that your newly released capabilities are generating, and if you are not feeling good about the return, then this is why changing how you solve problems is so important.

The root of the issue is that these feature teams are set up to *serve the stakeholders* in your business, rather than to *serve your customers in ways that work for your business.*

The business leaders and other stakeholders each understand their specific operational needs, and they come up with a list of features and projects that they believe will help them deliver on their obligations to the business. They hand these priorities down to the feature teams, which are then asked to provide a product roadmap with dates and deliverables.

So, why do so few of these features actually generate the hoped-for return?

Realize that each of these features is a *potential solution* to some underlying problem. It might be a customer problem—maybe customers can't figure out how to use your product effectively. Or it might be a company problem—maybe it costs too much for you to provision your product.

In the feature team model, the feature on the roadmap is designed by the designer on the feature team and then built by the engineers on the feature team. But whether that feature will provide any value to your customers or your company is the responsibility of whoever requested that feature on the roadmap.

This is usually a stakeholder, who may understand their own needs but doesn't have the intimate understanding of the enabling technology so critical to tech-powered products, and rarely has the benefit of the latest conversations with users and customers concerning their needs and issues.

As a result, in the feature team model, you can't hold the feature team accountable for business results. The team is there only to produce *output*—the *features*. If a feature doesn't generate the desired *outcome*, the team will simply point out the fact that they just built what they were told to.

Now, it's also very true that the stakeholder who decided on this feature will not want to take responsibility for the failure, either. They'll almost certainly argue that the feature that was shipped was not what they had hoped it would be or that it took longer to deliver than they had expected. Which is why lack of trust between stakeholders and feature teams is such a common complaint.

Another serious consequence of this way of working is that you end up with a great many "orphaned" features—features that don't produce real value but are waiting for the day they might get another iteration, which rarely happens. The result is a very fast accumulation of technical debt, which can quickly get out of hand and end up dramatically slowing down the teams. At its worst, it can bring the business to its knees.

For whatever reason, as soon as you lose your ability to consistently create value—to innovate on behalf of your customers—it is only a matter of time until your competitors are able to offer a more compelling solution to your customers. This has happened to countless companies.

You can reduce your prices, and you can run clever marketing and sales promotions, but these measures at best just delay the inevitable. Eventually, someone else will serve your customers in ways that you no longer can.

Empower Teams with Problems to Solve

Let's contrast this with an empowered product team. In this case, rather than providing the product team with a roadmap of features and projects to build, the product team is instead given *a set of problems to solve and desired outcomes to achieve.*

It is not difficult to sit down with the relevant stakeholders and work backward from the specific features they are requesting to what the underlying problem is for the customer or for the business, and then discuss how success would be measured and what the desired outcome is.

Rather than simply implementing the features desired by the stakeholders, an empowered product team is tasked with coming up with a solution that works for both the customer *and* the business.

This means coming up with a solution that is *valuable*—the customer will decide to buy or use it; *usable*—the user will be able to figure out how to use it; *feasible*—your engineers know how to solve with the time, skills, and technology on the team; and *viable*—the solution will work for your business in terms of constraints in marketing, sales, finance, service, legal, and compliance.

Why is an empowered product team necessarily any better than feature teams?

Besides obvious reasons, such as improved morale resulting from a much greater sense of ownership and the knowledge gained in testing potential solutions directly with your users and customers, the major reason is because strong product companies understand that *empowered engineers are absolutely essential to innovation*, and empowered product teams are designed to tap into this powerful asset.

In a feature team, the engineers are there simply to build what is requested, and the designers are there to make it look pretty. They are effectively *mercenaries*. Even worse, if you outsource your engineers, they are quite literally mercenaries.

But in an empowered product team, the engineers are not just there to build, and the designers are not just there to design. They are also responsible for helping to *identify the right solution*. This is what is meant by the term "empowered product teams."

The advantage that the engineers have is that they are working with the enabling technology every single day, which puts them in the ideal position to see what's *just now possible*.

When these empowered engineers work collaboratively with product managers and product designers, and are exposed directly to users and customers, you can start to see where nearly every innovative product or service you love originates from.

We are admittedly emphasizing the role of empowered engineers here especially hard, but that is because companies that have not yet transformed typically require the biggest philosophical shift in terms of how they think about engineers. Evidence of that fact is the degree of outsourcing of engineers versus the other roles.

When combined with a strong product designer trained in the craft of designing effective and engaging user experiences and a capable product manager trained in understanding both customers and the constraints of your business, you have the cross-functional set of skills necessary to solve hard problems in ways your customers love yet work for your business.

Product Discovery

While time to market is still important for an empowered product team held accountable to results, what is most important is *time to money*—in other words, the time to achieve the necessary *outcome*.

If an empowered product team ships a feature but does not see the necessary impact, then they iterate on that feature or on their approach until they do.

With time to money as the goal of the product team, the incentive moves to being able to quickly determine if a particular product idea or approach will work. This is referred to as *product discovery*.

The product team could simply build their various ideas and see what works, but this would take much more time and expose your customers to a lot of bad ideas. Instead, the product team has a full battery of tools and techniques to quickly test ideas and approaches.

Mostly, the product team creates *prototypes*, which come in various forms, but all are very fast and inexpensive to create. Each prototype is designed to test different risks or assumptions.

The general rule of thumb is that any idea tested in product discovery should be *at least* one order of magnitude less expensive and faster than using the engineers to build, test, and deploy an actual product. In many cases, the prototypes are two orders of magnitude less expensive than the product.

If, on average, it takes three to five iterations for a typical feature to reach the point where it generates the necessary business results (time to money), and if a feature team is being used to build each of these iterations, taking several months for each iteration, then you're often talking on the order of one to two years before a feature generates the desired returns. And that's *only* in the rare case the stakeholders are willing to continue to put these necessary iterations on the product roadmap.

In contrast, if the empowered product team is assigned the problem to solve and it's staffed with the skills needed to perform product discovery, then those three to five iterations can happen in a small number of days or weeks, and the product version is in the customer's hands—generating the necessary business results—usually just a few weeks after that.

If you've ever wondered how small, empowered product teams can so consistently outproduce much larger companies spending much more on their feature teams, this is right at the heart of the answer.

We discuss how to develop the skills necessary for changing how you solve problems in chapter 17, "Product Discovery."

True Collaboration with Stakeholders

Beyond saving money and time, the more important benefit of changing how you solve problems is that you are creating the mechanism to consistently create value for your customers.

Instead of feature teams there to *serve the stakeholders in your business*, you now have empowered product teams designed to *serve your customers, in ways that work for your business*.

This difference is not minor, and it fundamentally changes the dynamics of your organization.

But be warned:

Some key stakeholders will not be happy about losing control over the technology resources. It is likely that some will resist passively, and a few actively. This is why—without the visible and real support from the very top of the organization—transformations so often fail.

It's also important to point out that some of your current product managers, product designers, and engineers may not be willing or able to take on this additional responsibility.

In truth, it is much harder to take responsibility for solving customer problems than it is to just build the features you've been told to build. You will need to ensure that you have staffed the right level of people on your product teams and that they are provided with the necessary coaching and context for them to succeed. See part III, **Product Model Competencies**.

Strong, cross-functional, empowered product teams composed of product managers, product designers, and engineers—working together to solve customer problems in ways that your customers love yet work for your business—is the essential core concept of the product model. See chapter 15, "Product Teams."

Outcome-Based Roadmaps

If you're working on changing how you solve problems, this begs the question: Where are the problems coming from?

In prior models, problems usually are coming from stakeholders as they prioritize their needs in a product roadmap. In a company that's moved to the product model, these problems instead come from an insight-driven product strategy.

But what if you want to focus on building your skills in solving problems, before you tackle the product leadership skills of product strategy?

For this situation, there is a helpful transformation tactic called *outcome-based roadmaps*.

If you go to your stakeholders today and ask them for problems to solve, that's likely to cause confusion because most product roadmaps are actually a list of *outputs*. Rather than problems to solve, they are features to build.

Fortunately, it's not difficult to reverse engineer the problem to solve.

The way this works is that you start with your existing output-oriented product roadmap, but then you go through each item on that roadmap and determine what problem each feature is hoping to solve, and what the measure of success (expected outcome) is.

You will still be working on the same problems that you would have before transforming, but now you are providing your product teams with a degree of freedom as they explore potential solutions, which enables the teams to focus on outcomes. Then you can learn the techniques of product discovery to help you achieve those outcomes.

An additional benefit of outcome-based roadmaps is that they help wean your stakeholders off talking in terms of features and dates and instead discussing problems to solve and outcomes.

CHAPTER

9

Changing How You Decide Which Problems to Solve

So far we've talked about the need to change how you build and the need to change how you solve problems. But we haven't talked about how you decide *which* problems are the most important to solve.

It is not hard simply to take your existing product roadmaps and, for each feature or project, to determine what the underlying problem to solve is and what the logical way to measure success would be. That's the simple and straightforward way to move from roadmaps of features to roadmaps of outcomes. (This is referred to as an *outcome-based roadmap*—see the box in chapter 8.)

This is not a difficult step, and it's true that simply providing your empowered product teams with problems to solve—and clear measures of success—can go a long way toward generating significantly better solutions to those problems for your customers and your business.

But are those really the most important problems to be solved for your customers and for your company?

Every company has before it a set of opportunities and faces a set of threats.

How rigorous is your company in selecting the *best* opportunities to pursue and focusing on those threats that you should take seriously?

Disrupting Product Planning

In most stakeholder-driven, feature-team companies, there is some form of an annual product-planning process. Here, the stakeholders make their cases for the projects they think are most important to be done.

In many of these companies, the finance organization plays the role of arbiter and makes decisions based on submitted business cases.

In other companies, the stakeholders get in front of the senior leadership to plead their case, and the senior leadership team makes the decisions.

Yet, companies that genuinely care about accountability track the results delivered versus the promises made, and they know that neither of these approaches results in consistently good decisions, primarily because they don't have good data going in. The book *INSPIRED* goes into detail about why this is the case and what good product teams do instead.

Product planning is about deciding which problems you need to solve. Changing *how you decide which problems to solve* is the third dimension of moving to the product model.

Disrupting Yourselves Before You're Disrupted

It's very easy to overlook just how fundamental this change is, and not recognize that it is the essential ingredient in so many companies that have disrupted *themselves* before they've been disrupted by *others*.

Suppose you know that, due to competitive pressures and changes in customer needs and behaviors, you need a substantially new generation of your core product. But you also realize that the needed changes will run deep—impacting how your products are created, marketed, sold, delivered, and serviced.

No matter how many company offsites you hold or how many management consultants you bring in, do you *really* believe that, at the end of the day, each of the stakeholders in your company will do what's necessary to disrupt themselves?

And even if they wanted to, the odds that the answer is in the conference room they are sitting in is incredibly unlikely. It is very challenging to innovate meaningfully that far away from the technology and from the customer.

We'll guess you've probably realized that's very unlikely to happen.

The product model understands this reality. It's not that the product organization is given carte blanche to do what's necessary. But with the support of the CEO, the product leaders and product teams work with the stakeholders to drive the necessary changes across the organization.

You'll see this very dynamic play out in each of the three detailed transformation stories in this book—Trainline (part V), Datasite (part VII), and Adobe (part IX)—but let's start with a high-level description of what's happening.

Customer-Centric Product Vision

So many companies spend their time *reacting*—reacting to new sales opportunities, reacting to competitors' offerings, reacting to customer requests, reacting to price pressure, and more. Yet, strong product companies care about these factors, but they are not driven by them.

What drives these companies is the pursuit of a product vision that can meaningfully improve the lives of their customers.

In fact, in a strong product company, a compelling and inspiring product vision is your single best tool for recruiting the people on your product teams. The people you want on your teams are precisely those who believe in your product vision. They want to make a difference in the lives of your customers.

A strong product vision will inspire an organization for many years (most run three to ten years).

It's important to note that the product vision is, first and foremost, about the *customer*. How will it make the lives of your users and customers better?

It is *not* about how you are going to make more money, or what your priorities should be for the quarter, or how you'll structure your product teams. These other topics are important, and we discuss those goals next, but the purpose of the product vision is to *describe the future that you are trying to create*.

You may have dozens or even hundreds of product teams, but the product vision is what unites the teams with the shared goal of making that vision a reality.[1]

Insight-Driven Product Strategy

While the product vision describes the future, the product strategy is how you identify the most important problems to solve *now*.

The product strategy begins by focusing on the most critical areas for business success. The reality is that most companies try to do too many things at once. So, they end up diluting their efforts and making too little progress on the true levers for the business.

As legendary CEO Jim Barksdale said, *"The main thing is to keep the main thing the main thing."*

Or, if you prefer the Russian proverb: *"If you chase two rabbits, you will not catch either one."*

Once you have identified the few key focus areas, you then need to select the *insights* that you will be betting on.

You are constantly investigating the quantitative insights—mainly derived from the data that's generated—and also the qualitative insights—mainly resulting from talking directly to your customers. You're also continuously evaluating the potential of new enabling

---

[1] If you'd like to see some of our favorite examples of inspiring product visions, check out: www.svpg.com/examples.

technologies and the implications of relevant industry and technology trends.

This type of insight-driven product strategy usually requires developing new muscles for the company, but the reward for strong product strategy is to get the most out of your technology investment. *Good product strategy is a force multiplier.*

It's important to understand that product strategy is distinct from both *business* strategy and *go-to-market* strategy. Many companies have strong skills in business strategy and/or go-to-market strategy, but product strategy is often missing completely.

In fact, in most feature-team companies, the product strategy is literally to try to deliver as many features as possible for the different stakeholders. Which is to say, there really isn't a product strategy at all.

The Role of Product Leadership

Most companies that want to transform understand that they need to dial up the level of their product teams. Hiring more senior engineers. Hiring skilled product designers. Hiring competent product managers.

But many companies are surprised to learn that they often have even larger gaps in their product leadership.

In fact, many feature-team companies really don't have any product leadership to speak of. Since they rarely have a product vision, and are even less likely to have a product strategy, the job of a "head of product" in a feature-team company is completely different from the same title in a company operating in the product model.

More generally, in the product model, the product leaders play the primary role in coaching and developing their people to have the skills to discover and deliver solutions effectively and successfully.

In the product model, the leaders of product management, product design, and engineering are absolutely essential not only to changing how you decide which problems to solve, but also to changing how you solve problems and changing how you build.

We discuss how to develop the necessary skills in deciding what problems to solve in chapter 16, "Product Strategy."

The One Right Way?

A very common but damaging misconception in the broader product community is that many people try to argue there's only one right way to build great products.

Sometimes this confusion happens because a person or team will try out a method or technique or process and have a great result, so they think that all they have to do is replicate what they had done earlier.

More often, this confusion is motivated by people and businesses that just want to sell you their particular process, framework, methodology, or tool.

But the product operating model is a *conceptual model*.

It's a way for you to *think* about the work. It is not a prescription. It is not a process. There is no single framework or methodology or tool.

The reality of product work is that you have a tremendous range of problems to solve, each with different risk profiles, different types of customers, different types of technologies, and different constraints to satisfy.

Even if there was one approach that could succeed for everything, that approach would certainly be overkill—too slow and too expensive—for the vast majority of less complex work.

Building a consumer device is very different from building an AI-powered SaaS service. Even within a single product organization, there are often dramatically different product situations.

This is why we focus on the product model *principles*. The principles apply no matter the complexity, the customer, the technology, or the industry.

So, when you assess an organization, you need to look below the surface to understand what the context is, and whether the team is making appropriate choices.

There are examples of this confusion everywhere. Earlier we mentioned how many organizations follow Agile rituals and ceremonies, use Agile coaches, and staff the Agile roles, yet they only release monthly or even quarterly. Contrast that with so many strong product

(continued)

companies that don't see the need for Agile methods or roles or rituals, yet they have long practiced continuous delivery. It's not hard to see which organization is actually being true to the principles underlying Agile.

Another example has to do with the various product competencies. There are strong product companies that combine competencies. For example, it's not uncommon for a product designer to also be a front-end engineer. There are clear advantages to this, although the pool of candidates is significantly smaller. Yet the real question is not how these companies define the designer role; it is how they use those designers to minimize waste, address risks, and create valuable and usable products.

PART

Product Model
Competencies

In prior models, product teams exist to serve the needs of the business or, more accurately, the business leaders.

But in the product model, *product teams exist to solve hard problems for your customers and for your business, in ways your customers love, yet work for the business.*

This difference may sound minor, but the consequences are quite significant in terms of how teams work and how they interact with the rest of the company.

Consider for a moment the implications of this. We are essentially pushing the decisions and responsibility for finding the best solution to the problem down to the relevant product team, and then holding that team accountable for the results.

This is what drives the need for the new product core competencies.

Strong products are created by strong product teams, and strong product teams are staffed with the critical new product competencies.

In this section, we discuss each of these critical new core competencies so that you know what to look for and what to expect, and are not fooled when someone adopts a new title but doesn't have the necessary skills or experience to back it up.

It's important not to underestimate the effort to establish these new competencies. Everything that follows in this book is built upon these competencies.

To be very clear, unless you are willing to establish these new competencies, your transformation hopes will likely end here.

It's important to acknowledge that these new competencies can feel threatening to people who have built their careers around different competencies.

Some will view the product model as a major career opportunity and a chance to stay relevant for years to come. Others will dismiss the topic, thinking they already do this, and others will simply prefer to wait it out in hopes that the leaders will lose interest after a few months.

It's also important to emphasize that many of your existing people can be coached and developed into strong contributors with the necessary competencies. But don't expect that to be true of all of them.

And everything depends on your people *wanting* to learn and then having leaders or product coaches that are *capable* of helping your people learn these new competencies.

In this part, we describe each of the new core competencies, but not in enough detail that you can learn them.

Each of these specialized competencies normally takes years to learn. In fact, we have published two other books that attempt to teach the skills and techniques needed for these competencies.

Rather, we want to describe the new competencies such that *you understand the need for these competencies and you know what you can and should expect from people with these job titles.*

We also want you to start thinking now about changing your job descriptions and career ladders to reflect and secure these new competencies.

One of the challenges in transformation is that many people have adopted the new titles of the product model without actually *learning* the new competencies. This problem is much more severe than you might imagine.

We want to give you the tools to be able to judge competence effectively.

Product Managers, Product Designers, and Engineers

A typical cross-functional product team requires three specific and very distinct skill sets.

Usually that means at least three people. But sometimes a single person is able to cover multiple skill sets or, in certain specific cases, the particular product may only require two of the skill sets.

Recall that when solving problems for your customers or your business, you need to come up with solutions that your customers love, yet work for your business.

To discover an effective solution, the team is responsible for addressing four different types of risks:[1]

1. *Value risk.* Will the customer buy our solution or choose to use it?

2. *Viability risk.* Will this solution work for our business? Is it something we can effectively and legally market, sell, service, fund, and monetize?

3. *Usability risk.* Can users easily learn, use, and perceive the value of the solution?

4. *Feasibility risk.* Do we know how to build and scale this solution with the staff, time, technology, and data we have?

---

[1]There are other risk taxonomies besides the one we present in this book. The main alternative is from the product design firm IDEO, which breaks down the risks as: *desirability*, *feasibility*, and *viability*. In other words, IDEO combines value and usability risks into one risk termed *desirability*. The most important point is to use some kind of risk taxonomy to ensure you are considering all the important risks. That said, we have found that the IDEO taxonomy works better for consumer products (where every user is also the buyer) than it does for business products, as it is too easy to confuse and conflate whether a user *can use* a product versus whether a customer *will buy* a product.

In a cross-functional product team, these are the critical competencies and what each is responsible and accountable for:

- The **product manager** is responsible for the *value* and *viability* risks and is overall accountable for achieving the product's *outcomes*.
- The **product designer** is responsible for the *usability* risk and is overall accountable for the product's *experience*—every interaction your users and customers have with your product.
- The **tech lead** is responsible for the *feasibility* risk and is overall accountable for the product's *delivery*.

It's also important to realize that each of these three competencies can and do contribute to all aspects of the solution, but it's helpful to know who specifically is held accountable for each risk.

As you'll see, the skills required to develop each of these competencies are substantial.

Product Leaders

Product leaders are the managers of product management, product design, and engineering.

These are the people who will need to recruit, onboard, coach, and develop the actual product managers, product designers, and engineers.

In addition to building and coaching the product teams, these product leaders have critical responsibilities related to creating a compelling and inspiring product vision, an insight-driven product strategy, and a carefully crafted team topology, and to defining the critical problems to solve and business outcomes to achieve.

More generally, these leaders are responsible for ensuring that the product organization is aligned with the broader company in pursuing the best opportunities and addressing the most serious threats and in influencing the stakeholders/ecosystem around the product teams to make this change successfully.

This is why people you select as your product leaders will very likely prove critical to your transformation efforts.

If you don't yet have product leaders with the necessary skills and experience, then your first choice is to recruit them and get them in place for the kickoff of your transformation work.

That said, in many companies, product leaders have some but not all the necessary experience, and external product leadership coaches can temporarily supplement their knowledge.

Other Impacted Roles

Finally, it is important to point out that there are a few other roles that are significantly impacted by the move to the product model. One is product marketing and another is project management (aka delivery management). These and other roles are discussed later in the book.

CHAPTER

10

Product Managers

*W*e will warn you now that this is consistently the most difficult new competency to establish.

Your company very likely already has people with the title "product manager." However, if this is the case, it is also likely that these people were simply retitled from other roles, such as "product owner" or "business analyst" or "program manager."

The reason this competency is so messy is because your prior model very likely had a role with the same "product manager" title, but in the prior models, that is a completely different job, with very different skills and responsibilities.

The result is that when it comes to this title, you will need to look deeper to determine on a case-by-case basis which of your people truly have the necessary skills, and if not, whether you believe that with skilled coaching they will be able to achieve the necessary level of competency in a reasonable amount of time.

As discussed earlier, in the prior models, the stakeholders (if anyone) are implicitly responsible for the *value* and *viability* of the solutions that are built. But in the product model, the product manager is explicitly responsible for ensuring the value and viability.

Value requires a deep understanding of your users and customers, and *viability* requires a solid understanding of your business.

The book *INSPIRED* goes into depth on this role, so we won't try to duplicate that here, other than to make sure you are able to assess whether the people in this role will be able to do the job that is necessary.

Understanding your customers means not only being an expert on how your users and customers choose and use your products—both qualitatively and quantitatively—but also understanding the market, the competitive landscape, and the relevant technology and industry trends.

Understanding your business means learning how your product is funded, monetized, manufactured, marketed, sold, delivered, and serviced, as well as any legal, contractual, or compliance constraints.

It is normal for a new product manager to need on the order of two to three months to come up to speed on your customers and your business—and that's with the active coaching of a capable manager.

Without that active coaching from a strong manager, the person could easily be in the job for multiple years and still not have the necessary knowledge. The most common visible symptom of this situation is a lack of trust in that product manager by the company's stakeholders.

Realize that because normally the engineers and the product designer do not have this knowledge (although we love it when they do), if the product team is to be able to make good choices as they work to discover an effective solution, they need this knowledge to come from somewhere.

One clear sign that the product team lacks a competent product manager is that all decisions require either an escalation to a manager or a stakeholder meeting where the stakeholders are asked to decide.

To set your expectations about the necessary level of skill required for this competency, here are two candid realities:

1. Your head of product will be judged by her *weakest* product manager.

2. Your CEO should believe that each of your product managers has *the potential* to be a future leader of your company in the next five or so years.

While there are definitely exceptions, the majority of legacy business analysts and Agile-focused product owners simply do not reach this bar.

Those legacy jobs are very narrowly defined roles as compared to a product manager in the product model, with much less specific knowledge required.

Realize that your product managers will need to establish trust-based relationships with your major customers as well as with your key stakeholders. It takes strong people to do this.

Unfortunately, the reason why so many transformations are destined to failure is that the company is not willing to enforce this standard on the people they call product managers, so the product teams do not contain the necessary cross-functional skills, the stakeholders don't trust the teams, the product teams are not able to deliver the necessary outcomes, and the attempt to move to the product model collapses.

If you believe you can simply retitle your product owners or business analysts, you are very likely heading for failure.

While it is very possible that you currently have several people who will need to find other roles where they are better suited for success, it is equally likely that you currently have very strong candidates for the product manager role scattered across your organization.

A very important part of transformation is identifying these high-potential people, getting them into this critical role, and providing them the necessary coaching to succeed.

Look for people with a broad understanding of business and the proven ability to quickly learn other parts of the business. Look for people who are comfortable immersing in the data to understand behavior and trends. Look for people who want to get face-to-face with real users and customers. Look for people who are willing to roll up their sleeves around a whiteboard with designers and engineers to try to come up with solutions that solve for many different constraints.

Look for people who love to quickly learn as much as they can and collaborate to solve hard problems.

One final attempt at impressing upon you how important this point is: If you are thinking that because your company hires only good people, this isn't a problem for you, in our experience, you are on your way to a failed transformation.

Direct Access

For the product team to discover and deliver effective solutions, it is absolutely critical that they—especially the product manager—have *direct, unencumbered access to* these three constituencies:

1. Users and customers
2. Product data
3. Business stakeholders

Direct Access to Users and Customers

Without direct, unencumbered access to actual users and customers, the product team has little hope of any kind of success.

Users and customers provide not only inspiration for unaddressed problems but also the means to rapidly test your proposed solutions.

It should be intuitively clear why the product manager and product designer need direct access to users and customers, but less obvious is the importance of providing your engineers with direct access to the users.

It's important to acknowledge that you don't need all your engineers to tag along on every customer or user interaction. But "the magic happens" when an engineer sees a user struggling with their product, so the more you can encourage and facilitate this interaction, the better.

Of course, you need to be sure your people have been trained and coached on appropriate interactions with customers, but the key here is not to let anyone prevent product teams from having direct access to customers—not a sales or marketing person, not a customer success person, not a user researcher, not a customer's vendor manager. No one.[1]

That said, there's no reason you can't occasionally include others in your direct interactions with users. For example, it can be very impactful to include a key stakeholder, or an interested executive, especially when you see that shared empathy is needed.

Direct Access to Product Data

Similarly, the product manager needs direct access to the product data in order to make good decisions based on that data.

Usually several data sources contain different types of data—such as how your users are interacting with your product, how your customers purchase your products, and how your customers' behaviors are changing over time.

The product team may have access to data analysts and data scientists to help explain complex data patterns, but decisions based on this data are the responsibility of the product manager, so she needs direct data access.

In some companies, certain people want to restrict access to this data, and it is important that customer privacy and security be maintained.

You can create systems that enforce the appropriate types of governance, access, anonymization, and aggregation that may be required but still give teams direct access to essential data.

As with direct access to customers, direct access to data is fundamental to the product model. Without direct access to the data, the product manager is flying blind.

(continued)

---

[1] Most companies intuitively understand the need for product teams to engage directly with users and customers, but objections arise in some companies. We cover these in part X, **Overcoming Objections**. For now, understand that product teams directly interacting with customers on an ongoing basis is absolutely fundamental to the product model.

Direct Access to Business Stakeholders

Discovering a solution that works for your business is about solving for the various business constraints—marketing, sales, services, finance, legal, compliance, manufacturing, subject matter experts, and more. And it's essential that the product manager has direct access to the leaders across your business who are responsible for these constraints.

The product manager must establish relationships with these people, putting in real time and effort to learn the concerns and needs of the various stakeholders. Stakeholders must genuinely believe that the product manager understands the relevant constraints and will ensure they are considered and addressed in any proposed solution. And the product manager will discuss with stakeholders before building anything in that gray area where viability is not certain.

This trust between product manager and stakeholders is critical to the healthy collaboration that's needed to find effective solutions that solve for both the customer and the business.

Direct access to users and customers, direct access to data, and direct access to stakeholders is key.

Direct, ongoing access to these three constituencies is what enables the product manager to ensure the value and viability of the solutions that are built.

If your product managers are prevented from accessing any of these three constituencies, then they are very likely going to fail.

What all this means is that you need to ensure the product team's direct access to these three constituencies and fight any attempt to place a well-meaning person, or cumbersome process, between the product manager and these constituencies.

In some companies, especially those with a legacy of outsourcing engineers, the product manager needs direct, unencumbered access to an additional group: the engineers.

This should be the most obvious statement in the world, but it is not obvious to many people coming from the IT model. Access to engineers may not be a given for those working in environments where. . .

(continued)

- The engineers are outsourced
- The engineers are located in time zones where interaction is difficult, or it is believed that engineers need to be sheltered so they aren't interrupted while they code
- The product team is split up such that the product manager interacts with customers and stakeholders, and some other person—such as a product owner or project manager—interacts with the engineers

Because the engineers are working daily with the enabling technology, they are in the ideal position to understand what's *just now possible*. But this only happens in a way that is actionable when engineers are interacting with the product manager every day.

At an absolute minimum, the product manager needs direct access to the tech lead.

What About Domain Expertise?

You may have noticed that one attribute noticeably missing from the description of a strong product manager is *domain expertise*.

For example, if your product is an insurance product, just how much do you need to know about insurance?

The reason this is a such an important topic is because it is so common for companies that are transforming to hire product managers out of their customer base or others with deep domain experience.

However, in most cases, this turns out to be a mistake. It is likely no surprise that someone who knows too little about the domain will struggle to do the job. What is much more surprising is the problems that are caused when a product manager knows *too much* about the domain.

To be clear, we expect new product managers to become expert in their domain in their first months of onboarding in the role. This is

(continued)

(continued)

about the dangers of hiring people *because* of their preexisting domain experience.

Product thought leader Shreyas Doshi refers to true domain expertise as domain *knowledge* minus domain *dogma*. For so many people who spend many years immersed in the domain, it is very difficult to distinguish dogma from expertise. The problem is that the domain dogma seriously impedes innovation. These are the people who can recite all the reasons something *won't* work, until someone comes along without that dogma and shows that it *can* work.

For certain domains where deep domain expertise is required, we will hire a dedicated domain expert who can be shared by the various product teams, such as surgical instruments, tax preparation, or regulatory reporting.

11

Product Designers

M ost companies transforming to the product model already have
some number of designers, but they typically play a very different role than what's needed in the product model.

In the prior models, designers too often play a service role to a
product manager, or they support the marketing organization, none of
which is what we need.

In terms of what we need in the product model, there are really
two issues here: how the product works, and product discovery.

How the Product Works

The first issue is that in the prior models, the designers often have
a much smaller range of skills. They are usually graphic or visual
designers, and while those are definitely helpful skills, as Steve Jobs
famously said, *"Design is not just what it looks like and feels like. Design is
how it works."*

In the product model, we need designers to take on a much larger role, responsible for the holistic customer experience and how the user and customer experience the product's value.

In the product model, we refer to the designers as *product designers*, and we need designers who are skilled in service design, interaction design, visual design, and, in the case of devices, industrial design.

We don't need product designers to be expert in *all* these forms of design, but for most products, it is critical that the product designer has proven interaction design knowledge and skills.

Interaction design is about understanding how technology and people (users and customers) communicate with each other.

Product Discovery

The second issue is that your current designers generally are brought in at the end to "make things pretty" or at least to apply a veneer so that it appears your products all come from the same company.

Instead, you need product designers who can help *discover an effective solution*.

Skilled product designers are trained to engage directly with users and customers to get inside their heads—to understand how they think about the problem you are working to solve for them and how they think about your potential solutions. They strive to design *holistic experiences* that will feel recognizable and intuitive to your users and customers.

Moreover, skilled product designers not only design experiences, they intentionally design *changes to those experiences*.

When the changes are designed well, users don't even notice the constant stream of changes and improvements, but when users do notice them, the experience guides them through the changes rather than requiring any form of retraining or disruption.

The product designer is the key partner for the product manager and the tech lead in the daily product discovery work. The product designer usually takes the lead in expressing the product ideas in the form of prototypes that you can experience, consider, evaluate, and test.

It's not uncommon for a typical designer to create a dozen or more prototypes in a week. Although most of them will get thrown away, every one of them helps you converge on an effective solution.

To set your expectations, due to the level of skill required and the market demand for product designers at companies following the product model, these highly valued professionals are generally compensated at the level of a product manager and an engineer.

Normally, each product team that is responsible for user-facing products or services would have a dedicated product designer.

As more companies move to the product model, and as more types of product companies—such as those that sell to large businesses—realize they need the help of product designers, demand for these professionals continues to grow.

12

Tech Leads

There is an important difference when discussing engineers versus product managers or product designers.

With product managers and product designers, you only have one dedicated person per cross-functional product team. So, there are quite a few responsibilities and expectations for each of those people.

However, there are usually multiple engineers on each product team.

Different types of product teams—working with different technologies and with different scopes of responsibility—will need a larger or smaller number of engineers with a range of specialized knowledge.

But because you have multiple engineers, you can have a mix of skills and experience levels. For example, on a given product team, it is normal to have at least one senior engineer covering the tech lead role, some engineers with a few years of experience, and maybe a couple of engineers who are just starting their careers.

Some companies prefer their engineers to have more experience, and others prefer those just starting their careers so they can coach good habits from the start. This choice is up to the head of engineering.

What we need to discuss here is the role of the most senior engineers, which we refer to here as *tech leads*.[1] To be clear, the tech lead is normally an individual contributor and not an engineering manager.

There are two dimensions when we talk about the difference between engineers in the product model versus engineers in the prior models.

The first dimension has to do with the complexity of the solutions that product companies build versus typical "IT systems."

It's important to acknowledge that nearly all technology looks easier until you actually get deep into it. But that said, there are some fairly significant differences around scale, performance, fault tolerance, reliability, internationalization, test automation, deployment infrastructure, new emerging technologies, and architecture and strategies for managing technical debt that distinguish product engineers. Experience with these techniques is extremely valuable, as mistakes in these areas can be very expensive.

The second dimension has to do with whether the engineers are there simply to implement the solution that the product manager and product designer come up with or to help determine what the most effective solution is (and then implement that solution).

Of all the principles underlying the product model, the single most important is the realization that innovation absolutely depends on empowered engineers.

That's because the engineers are working with the enabling technology every day, which puts them in the best position to see what's just now possible.

But that's *only* true if the engineers care just as much about *what they build* as *how they build it*.

The old and misguided adage that "a PM is responsible for the *what* and *why*, and the engineer is responsible for the *how*" completely misses the point of empowered product teams and the source of real innovation.

---

[1] In this book we use the term "tech lead" for the most senior individual contributor engineer on the team, but many different terms are used for this key role. The title you use is not important. The key is that the mandate to tech leads is to care just as much about *what* is built as *how* it is built.

The primary reason for not outsourcing engineers in the product model is because they would be brought in far too late and would possess far too little understanding of the context or the customers to have any real chance at innovation.

In an ideal world, all your engineers would be engaged in determining what the best solution is, not just engaged in building that solution.

That said, it's not unusual for some of the engineers, especially the more junior engineers, to only care about building.

You can succeed so long as *at least the tech lead* on each team is willing to partner with the product manager and product designer to discover an effective solution to the problem the team is working to solve.

To be explicit on this critical point, *if the tech lead is unable or unwilling to engage in product discovery, then you are very likely guaranteeing that the eventual product will not achieve your goals.*

It really is essential that you establish this as a formal part of the job description of the tech lead.

In practice, this doesn't require much of the tech lead's time each day, but it does require some (usually less than an hour a day, every day, of the lead engineer).[2]

Most tech leads will tell you that a few minutes weighing in on a product idea very early can save weeks or even months of damage control later if decisions are made without consulting with them.

---

[2]Occasionally, we'll encounter an engineering team that has the opposite problem—they want to spend all their time in product discovery, usually because they are tired of just implementing what they consider to be an endless list of unhelpful features. You may need to remind these engineers that their primary responsibility is to build the products in delivery. But if you do encounter this situation, it's a good problem to have, and it's easy to correct.

CHAPTER
13

Product Leaders

So many people naively believe that the key to empowering product teams and the product model is simply to get the managers to back off, stop micromanaging, and give their product teams some space to do their jobs.

But in the product model, empowered product teams depend on *better* leadership, not *less* leadership.

What does that really mean?

As legendary CEO Andy Grove said, *"What gets in the way of good work? There are only two possibilities. The first is that people don't know how to do good work. The second is that they know how, but they aren't motivated."*

Let's tackle these two in order.

Management

Let's first discuss the management responsibilities, which primarily involve coaching and staffing.

Coaching

Probably the most often overlooked element to strong management in the product model is coaching. It is the single most important responsibility of every people manager to develop the skills of their people.

This most definitely does *not* mean micromanaging them. It does mean assessing and understanding their strengths and weaknesses, coming up with a coaching plan, and then spending the quality time necessary to help them improve.

More generally, every member of a product team deserves to have someone who is committed to helping them get better at their craft. This is why, in the vast majority of strong tech product organizations, the engineers report to experienced engineering managers, the designers report to experienced design managers, and the product managers report to proven managers of product management.

The amount of time and effort you need to spend on coaching depends on the number and experience level of your people. But to set your expectations, it's normal for first-level managers to need to spend up to 80 percent of their workweek on staffing and coaching.

Staffing

The managers are the people you hold responsible for staffing the product teams. This means sourcing, recruiting, interviewing, onboarding, evaluating, promoting, and, when necessary, replacing the members of the teams.

If you have an HR function at your company, they are there to support your managers with these activities, but they are in no way a substitute for the hiring manager in these responsibilities. This point is critically important for your managers to understand.

Because empowered product teams are predicated on competent product managers, product designers, and engineers, this starts with raising the bar on staffing and coaching.

Taking staffing seriously is hard, and it takes a substantial amount of time and effort. You'll likely feel like this is not the product-related work you prefer to do.

But as Jeff Bezos—Amazon founder and executive chairman—says, *"Setting the bar high in our approach to hiring has been, and will be, the single most important element of Amazon's success."*

Leadership

There are essentially two main ways you can lead a product organization.

You can lead by what's known as *command and control*, which means explicitly telling your people what you need them to do, usually by assigning them a roadmap of features and projects to build. In this model, the leaders and stakeholders make most of the meaningful decisions, and your product teams (or, more accurately, feature teams) are there to carry out those decisions. Admittedly, command and control is easier to do.

The alternative is that you can lead by empowering the teams. You assign them business or customer problems to solve and then let the product teams determine the best way to solve those problems.

However, if you choose to push key decisions down to the product teams, then you will need to provide those teams with the strategic context—especially the product vision and product strategy—necessary for them to make good decisions.

If a leader finds herself disagreeing with where the empowered teams end up, or swooping in toward the end to change direction, it is worth some reflection on what context the leader could have given up front to make the feedback on the back end unnecessary. This is a new skill that requires reflection and improvement.

This is why, for example, at Netflix the mantra is *"Lead with context, not control."*

The product leaders are specifically responsible for the product vision, the team topology, the product strategy, and the specific team objectives.

Individuals on the product teams may contribute ideas or insights, and that's a great sign of a strong culture, but these are ultimately leadership responsibilities as they span *all* product teams.

Product Vision

The product vision describes the future you are trying to create, and, most important, how the vision improves the lives of your customers. The product vision serves as the shared goal for the product organization. In terms of timeframe, it is usually between three and ten years out.

There may be any number of cross-functional, empowered product teams—ranging from a few in a startup, to hundreds in a large enterprise—but they all need to head in the same direction and contribute in their own way to making the product vision a reality.

Think about the product vision as a higher-altitude place to ensure alignment across the product teams and the stakeholders. If we can get alignment at this level, it will save a lot of debating about solutions later. Bringing everyone along also helps create empathy for users and customers.

Some companies refer to the product vision as their North Star—in the sense that no matter what product team you're on and whatever specific problem you're trying to solve, you always know how your piece contributes to the more meaningful whole.

More generally, the product vision is what keeps teams inspired and excited to come to work each day—month after month, year after year. It lets the teams proactively guide the product forward, rather than constantly reacting.

It is worth noting that the product vision is typically the single most powerful recruiting tool for strong product people.

Creating a compelling product vision is a bit different than the other elements of the strategic context. The product vision is more art than science. Its purpose is to *persuade and inspire*—it is meant to be *emotional*. You are talking about how you will improve the lives of your customers.

You don't want so much detail that your teams think it's prescriptive, but you also need enough detail so people can really understand what you're trying to accomplish.

While creating a good product vision is not easy, it is worth the effort because a good product vision is the gift that keeps on giving.

So much of what you do derives from the product vision—the architecture, the team topology, the product strategy, and of course, your products for the next several years.

Team Topology

The *team topology* refers to how you define the responsibilities and ownership of your different product teams. This includes the structure and scope of teams, and their relationship to one another.

Many companies have an existing team topology that was not intentional; it is simply a reflection of their organization (this common topology anti-pattern is known as Conway's Law).

Your goal with the topology is to maximize empowerment. You do that by striving for teams that are loosely coupled but highly aligned.

Coming up with an effective team topology is one of the most difficult, but most important, responsibilities for product leaders, especially at scale. It requires intense collaboration and negotiation between the head of product, the head of design, and the head of engineering. The decisions you make impact the relationships and dependencies between the teams and what each team will actually own.

When done well, your product teams are empowered with a high degree of autonomy, and they feel a real sense of ownership over their work and how it contributes to the greater whole. The teams can tackle hard problems, moving fast and seeing the results.

Product Strategy

The *product strategy* describes how you plan to accomplish the product vision, while meeting the needs of the business as you go. The strategy starts with focus, then leverages insights, converts these insights into action, and finally manages the work through to completion. We dive into what all this means in chapter 16, "Product Strategy."

More generally, the product strategy helps you get the most value out of whatever number of product teams you have.

The output of the product strategy is a set of business or customer problems to solve (team objectives) that the leaders will then need to assign to specific product teams.

The product strategy is where strong product leaders distinguish themselves.

They decide what the focus will be and what it won't be, and sometimes these decisions are not popular with other leaders. Strong product leaders live and breathe the data and insights about the product and constantly seek the points of leverage that power the product strategy. A strong product strategy can help a small organization outperform much larger competitors.

Unfortunately, there are no easy shortcuts to a strong product strategy. It requires real time and effort to aggregate and assimilate the data and insights you'll need.

Team Objectives

To execute on the product strategy, the leaders need to ensure that each product team has one or two clear objectives they have been assigned (typically quarterly), which spell out the problems they are being asked to solve.

These objectives derive directly from the product strategy—it's where insights are turned into action.

This is also where empowerment becomes real and not just a buzzword.

The team is given a small number of significant problems to solve—the *team objectives*. The team then considers the problems and proposes clear measures of success (the *key results*), which they then discuss with their leaders. The leaders may need to iterate with their teams and others to try to get as much coverage as possible of the broader organization's objectives.

The litmus test for empowerment is that the product team is able to decide the best way to solve the problems they have been assigned (their team objectives).

It takes strong leaders to be self-confident and secure enough to truly empower the people who work for them and to stand back and let teams take credit for their successes.

Ongoing Evangelism

The final critical role of the product leaders is communicating the product model and the strategic context—the product vision, team topology, and product strategy—both to the product organization and across the company more broadly.

This requires an ongoing crusade of evangelizing—in recruiting, onboarding, weekly 1:1 coaching, all-hands meetings, team lunches, board meetings, customer briefings, and everything in between.

The larger the organization, the more essential it is to be relentless at evangelism, and it's important for the leaders to understand that evangelism is something that is never finished. It needs to be constant.

As you can see, the job of a product leader in a strong product company based on empowered product teams is both very different from the role in the prior models, and much more difficult.

You will also notice that in each of the upcoming successful transformation case studies, the CEO took special care to ensure strong leaders of product management, product design, and engineering.

Product Ops

Product teams are in the business of trying out ideas quickly to determine if they work—before they spend the time and money to build a production-quality solution.

Sometimes you make decisions based on data you collect, and sometimes you make the decisions based on talking with actual users and customers about their experiences using your products.

Several companies have realized that, with a relatively small investment, they can increase the speed and effectiveness of their product teams by better supporting how those product teams make decisions.

User researchers are experts at preparing for direct testing with users and customers, running those tests, and drawing the right conclusions from the results.

(continued)

Data analysts are experts at helping you prepare live data tests, collect the necessary amount of data to achieve a sufficient level of confidence, and draw the right conclusions from the results.

Some companies build out large groups of user researchers and data analysts to run these tests *for* the product teams. While this might sound efficient, in such cases the product teams lack the benefit of the direct, intense, and very motivating interaction with customers and the results. And if the insights don't make it into the hearts and minds of the people on the product teams, the value of the insights is largely lost.[1]

Other companies leave all user and live-data testing to product teams, which also can work. But the teams usually are not as knowledgeable of the best techniques and methods and sometimes they can draw the wrong conclusions—resulting in additional time, iteration, and testing.

We have found what works best is to establish a small group of user researchers and data analysts who can coach and support a much larger group of product teams in moving quickly and making good decisions and in ensuring that insights are shared directly with product leaders, as these insights may benefit the next iteration of the product strategy.

In many companies, these people live in different groups, but in some companies, these people are brought together under a small **product ops** function, which can be helpful in bringing visibility to this important supporting function.

That said, you need to be very careful you don't set up any intermediaries, even when they have the best of intentions.

As discussed, for a cross-functional product team to be effective and successful, it needs to have *direct access* to users and customers, direct access to the data those users generate, and direct access to the various stakeholders across the business.

One note of warning:

Some companies define *product ops* very differently from how we've defined it here. The most serious issue is when a well-intentioned product ops group thinks it's their job to get in between the product teams and their users and customers, their data, or their stakeholders. Be careful that's not the case for you.

(continued)

---

[1] A recap of user research is about as useful as a recap of a vacation. The value comes in experiencing it.

Another danger that we discuss in more detail later is when the title "product ops" is used as a cover for a reconstituted program management office (PMO). This is a problem because old-style PMOs are often the manifestation of the command-and-control culture we're trying to replace. As you'll see, the prior model doesn't usually go away without a fight. So, you'll want to be aware of that possibility as well.

The Impact of AI on Product Teams

Newly developed AI technologies have enabled product teams to deliver a new generation of products and services able to solve existing problems in ways better than we have ever been able to solve them before, and these same technologies have also impacted the core competencies themselves.

As with so many other jobs, product managers, product designers, and engineers are all benefiting from significant productivity improvements brought by tools that help automate many of the more tedious and mundane tasks, and more generally help research and evaluate new ideas. And we expect this to continue, especially in terms of tools for the engineers.

But it is also true that, in many cases, there is more work to be done in terms of the ethical and business-viability risks.

The result is that today, product teams are able to do more than ever, but in many cases the stakes are higher.

As AI technologies and tools evolve, companies using the product model will continue to experiment with them—embracing and adopting what works.

Just as you continuously evaluate how new technologies can be incorporated into your products to help your customers, strong product teams will continue to evaluate how new technologies can help you produce these products.

14

Innovation Story: Almosafer

Marty's Note: This is an example of impressive innovation from a region of the world not known for these types of technology-powered innovations. But one of the things I've learned over the years is that talent can be found all over the world, and when these people move to the product model, they are capable of remarkable levels of innovation. Strong leadership combined with strong product teams were able to rise to the challenges posed by the pandemic.

Company Background

Almosafer (part of Seera Group) is a travel and tourism company based in the Kingdom of Saudi Arabia (KSA) that now serves the wider Middle Eastern region. The company specializes in meeting local needs of the region's countries and cultures and also tourism and religious travel into the region from all over the world.

The company began as brick-and-mortar travel agencies, and for many years these agencies served the personal, business, and government travel needs of the region.

But as in the rest of the world, after the internet arrived, soon there were several very large, global travel providers, and many brick-and-mortar agents were struggling to compete.

Those global travel providers offered the benefits of economies of scale, and technology-powered solutions, but they did very little to accommodate local needs, customs, and behaviors.

In many regions of the world, there was a clear opportunity for a regional player to emerge that both understood local needs and also had the skills and technology to address those needs.

By 2018, product leader Ronnie Varghese and technology leader Qais Amori—along with the other senior leaders of Almosafer—decided that to truly serve the customers in their region, they would need to build a digital team with the skills capable of discovering and delivering solutions that were superior to the global travel services. So, they set about transforming to the product model.

This transformation work began in earnest in 2018, and soon this work was delivering real results for their business.

Most important, they were building several new and essential muscles. They were now able to continuously interview customers to identify unmet needs, collect and analyze data to understand precisely what was working—and what wasn't—and quickly prototype and test out ideas. They were now able to discover solutions that their customers truly loved because they met their particular needs in ways that the major players simply didn't understand.

During this first year, the company concentrated on two major areas: significantly improving the customer experience and putting in place the technology infrastructure and architecture that would not only reduce its operational costs, but enable the company to innovate and serve their customers well into the future.

In just the airline flights side of the travel business alone, the company was able to use these investments to grow to a billion-dollar business.

Beyond the improvements in products, the digital team also built a reputation in the region as a different kind of company—one powered by a unique blend of talent and technology working like the best anywhere, and the company was proving to their community, their country, and the Middle East region that it could compete with anyone.

The group was well on its way to creating one of the true regional success stories.

Then the borders closed.

The Problem to Solve

As with the rest of the world, the COVID-19 pandemic largely brought travel to a halt. The KSA was hit especially hard because so much of its travel is international.

Fortunately, the previous year had prepared the digital team to be able to solve hard problems, although the pandemic certainly would be a test.

The teams had identified many potential opportunities through their continuous interaction with their customers—consumers and businesses. One such opportunity that they had observed earlier—but one that, with the pandemic, turned from an interesting possibility into an essential need—had to do with figuring out ways for family and friends to gather together safely.

In Saudi Arabia, there had long been a cultural tradition of *istiraha*, which translates to "resting place," but in general means a large, open courtyard or other outdoor area with room for grilling foods, sitting in the cool shade, and breathing fresh air. Families and friends would gather at an istiraha simply to be together or to celebrate special occasions.

With the pandemic, and the need to avoid indoor gatherings, finding and reserving these istirahas quickly became a serious and unmet need. Unlike with hotels or airlines, there wasn't even any form of registry or listing of these unique accommodations.

And as just one important difference, unlike lodging, where people reserve for the night, an istiraha is reserved from the afternoon to the evening of the same day.

Discovering the Solution

The team knew this would be an especially risky type of product because the vast majority of products are new solutions to existing problems; this was a new solution to a *new* problem (the need to find and book istirahas).

This required not just a strong solution, but the market also needed to be developed—essentially, the region had to be educated about this new type of offering in the marketplace.

In offering a new type of booking in the travel marketplace, the product team would need to work closely with business development and marketing to identify the istirahas that would be available for rental, so that they could ensure the supply, while at the same time discover the necessary customer experience for those looking to book istirahas.

The product team worked with product coach Hope Gurion (profiled later in this book) on this effort to ensure they quickly converged on a strong solution.

Another challenge was that the product team needed to adjust their work to operate remotely. Because they had built their skills and their trust within the team, they were able to transition quickly and effectively to remote collaboration and remote research with customers—both problem and solution discovery.

The Results

There were considerable unknowns with this new product. Nobody even knew just how many istirahas existed, let alone how many owners would be willing to list in the marketplace and then how many people would book.

The team's initial business goal was to demonstrate product-market fit for this new marketplace offering, which they quickly established. After that, the focus was on growth, and istirahas are now an integral and rapidly growing category of the marketplace.

Even beyond the financial success, this capability provided a real service to their country and region in a time of need, and few in the region have the illusion that any of the global travel companies would have cared enough to have invested anything.

Today, Almosafer has a truly remarkable 70+ percent market share in the region.

Such an extraordinary result is never from any single capability or offering. It is the result of a sustained effort of talent and technology to truly care for customers.

Almosafer is one of the true success stories in the Middle East, and the company continues to expand to more countries in the region. They are well positioned to continue their growth and success.

IV

Product Model Concepts

Moving to the product model involves not just several new competencies, but also several critical concepts—essentially, new activities to help you consistently create effective products.

Some of these concepts will resemble activities you've done in the past—although with very important differences—and others very likely will be completely new.

This section is meant to explain these critical product model concepts.

It's important to acknowledge that not everything involved in a transformation to the product model is equally important. A few concepts are absolutely critical, and those are the ones we discuss in the chapters that follow.

Others are better characterized as helpful but not likely to make the difference between success and failure.

Moreover, usually there are several possible paths to accomplish the same goals.

That said, for the critical product model concepts, one way or another, *you need to get them right.*

Get these critical concepts right, and you are very likely on your way to success. Get any of these critical concepts wrong, and things start to collapse.

Product Model First Principles

Each of the critical concepts we describe is based on a set of *product model first principles*.

These product model first principles represent the common beliefs in all the strong product companies we know, no matter whether they were born into the product model or if they have transformed. No matter if they are small or large. No matter if they serve consumers or businesses. No matter if they build software or devices.

Moreover, if you understand these product model first principles, you can quickly judge whether a new process, or a new technique, or a new role—or even the experience of a prospective team member you are interviewing—is helpful or potentially harmful.

Product Model Concepts

At the very foundation of the product model is the cross-functional, empowered **product team**. Most of what we discuss in this book is intended to help enable these product teams to do their work.

Since no company has as many people or product teams as it wishes, you need to be smart about getting the most out of the people you have. Doing this means having an inspiring product vision combined with an insight-driven **product strategy**.

In prior models, product strategy is much less relevant, as feature teams exist primarily to serve the needs of the business stakeholders.

However, in the product model, product strategy plays a central and critical role in identifying the best product opportunities, as well as the most serious threats, to select the most critical problems to be solved.

Once product teams have been assigned critical problems to solve, these product teams need to be skilled at **product discovery**, which is how you discover solutions worth building. You do that by assessing product risks before building anything, and then testing potential solutions in the fastest, least expensive ways you can.

Moreover, in the product model, working hard and discovering and delivering lots of features is not enough. You need to *solve problems* for your customers, in ways that work for your business. This means delivering real product *results*.

Once you believe you have identified a solution worth building, you need to build and deliver that solution to your customers. Doing this consistently, quickly, and reliably requires a set of skills referred to as **product delivery**.

To truly take care of your customers—both in responding to their needs and consistently delivering new value—product teams need to be able to deliver *frequent, small, reliable,* and *decoupled* releases that are instrumented and monitored.

Finally, an additional set of product model first principles contributes specifically to creating a strong **product culture** of consistent innovation.

One final but important point to understand: It takes a lot of work to establish this new product culture, but it takes very little to destroy it. You need to constantly remind yourself of the importance of the product model principles. There are always forces in a company that work to pull you off course. Protecting against this starts with constant awareness of these principles.

15

Product Teams

The most fundamental of all product concepts is the notion of an *empowered, cross-functional product team*. This is where effective, innovative products come from, and most of the product model is one way or another about creating and nurturing these product teams.

Principle: Empowered with Problems to Solve

First, let's look at what it means to be *empowered*.

An empowered product team exists to *solve problems* in ways that *your customers love, yet work for your business.*

What makes the team empowered is that they are assigned *problems to solve*—they might be customer problems or company problems—and it is their job to come up with the best solution to those problems.

It's important to recognize that this is very different from the feature teams that exist in the IT model, which are given prioritized lists of features and projects to build.

Second, let's look at what it means to be *cross-functional*.

A cross-functional product team has members who can cover each of the product model competencies necessary to come up with strong solutions to the problems you've been asked to solve.

It is so critical for an empowered product team to be truly cross-functional because it is not reasonable to expect a team to solve problems and to be held accountable to the results without the range of necessary skills.

Normally, that means a product manager, a product designer, and a set of engineers.

We've said this earlier, but it's so essential it bears repeating:

- The **product manager** is responsible for the value and viability risks and is overall accountable for the product's outcomes.

- The **product designer** is responsible for the usability risk and is overall accountable for the product's experience—every interaction your users and customers have with your product.

- The **tech lead** is responsible for the feasibility risk and is overall accountable for the product's delivery.

Some people like to refer to this three-person subset of a product team as the *product triad* or *troika*.[1]

It's normal that there are additional engineers on the product team as necessary, some with specific expertise, such as data science, mobile technology, or test automation/QA, and some who have a broad skill set able to work on many different technologies (often referred to as a *full-stack engineer*).

---

[1]While we don't mind the term "product triad," we don't normally use it ourselves for two reasons: First, sometimes the three core competencies are covered by two or four people, so it's not always three. Second, in our favorite product teams, all the engineers participate in discovery activities, not just the tech lead.

Product Leaders and Empowered Teams

An empowered, cross-functional product team is an incredibly powerful tool. Pick your favorite tech-powered product, and very likely there was an empowered product team behind it.

But it's important to realize that this kind of product team doesn't happen on its own.

In the product model, you depend on *product leaders to coach the members of these product teams, and to provide them with the strategic context necessary for them to make good decisions.*

In this book, we use the term "product leaders" to represent the managers and leaders of product management, product design, and engineering.

It's absolutely essential to understand that most people don't arrive at your company with the skills and knowledge they'll need to succeed as a member of an empowered product team.

Even great people coming from strong product companies won't know your strategic context.

So, it's the primary job of your product leaders to coach and develop the members of these product teams to get each person to the point where they are capable of doing the job they need to do.

Strong managers and leaders also actively work to remove obstacles impacting their product teams.

Principle: Outcomes over Output

At the end of the day, you know that unless the customer believes your new solution solves their problem sufficiently better than what they've been using, you fail.

Shipping lots of features may make you feel good, but unless that translates into real business results, you fail.

There are many variations on this same theme, including "accountable for business results" and "time to money over time to market."

They all speak to this principle that product teams exist to effectively solve problems for the customer and for your business, not just to build things that don't get bought or used.

As an example of the different behaviors this principle can drive, sometimes a team looking hard at the data will find that the best way to improve outcomes is actually to *remove* functionality. This is most common on mobile apps, where the amount of real estate is very limited and where everything on the screen is essentially competing with everything else for the user's attention. A team focused on outcomes will at least consider this possibility.

Principle: Sense of Ownership

If you want the product team to truly feel empowered and accountable, then you need them to feel a real sense of ownership over what they're responsible for.

Let's look at the scope of responsibility of this product team. In other words, *what is it that they own?*

The division of teams and their respective areas of ownership is the critical product leadership topic of *team topology*, but for now, you need each product team to be responsible for something *meaningful*.

It may be a whole product, but more often today it is some meaningful subset of a much larger product. It is not unusual for a single product or service to have dozens or even hundreds of product teams working on different aspects or components of the larger product, as perceived by the end user.

The product team needs to be responsible both for coming up with the solutions to the problems you've been asked to solve (referred to as *product discovery*) and building and delivering the solutions to your customers (referred to as *product delivery*).

Separating these two responsibilities into two different teams breaks this principle and causes serious practical and cultural problems. It's therefore critical that this sense of ownership applies to everything that is done—including discovery and delivery, major innovation work and minor optimization work, fixing bugs, and scaling users.

Note that this doesn't mean that every member of the product team spends equal time on discovery and delivery. In practice, product managers and product designers spend most of their day on product discovery work, and engineers spend most of their day on delivery work.

Principle: Collaboration

Collaboration is one of those words that is used so often, and in so many different ways, that it has lost its meaning for many people.

Of course, most people think they're collaborative.

But in the context of an empowered, cross-functional product team, being collaborative has a very specific meaning, and it is most definitely *not* how most people—especially most product managers—are inclined to work.

Too many product teams still follow the old waterfall process where a product manager defines requirements, hands them off to a designer to come up with a design that meets those requirements, and then hands that off—usually at sprint planning—to engineers who implement the requirements and design.

To be very clear, this is most definitely *not* what we mean by collaboration.

First, collaboration is *not* consensus. While you like it when the product team is in agreement on the best course of action, you do not insist on this. This is why you practice *disagree and commit*.

Similarly, collaboration is *not* democracy. You don't vote on decisions. Instead, you depend on the expertise of each member of the product team.

Generally speaking, if the decision pertains to the technology, you defer to the tech lead. If the decision pertains to the customer experience, you defer to the product designer. If the decision pertains to the business constraints, you defer to the product manager.

Occasionally there will be conflicts, and normally you'll run a test to resolve them.

Further, collaboration is *not* about artifacts. Many product managers think their job is to produce a document capturing "requirements," or at the least, they are there to write user stories. It is true that you often need to create artifacts (especially when team members are working remotely), but that is certainly not how you collaborate. In fact, these artifacts more often get in the way of true collaboration.

Why is that? Because once the product manager has claimed something is a "requirement," that pretty much ends the conversation and moves the discussion to implementation.

At this point, the product designer feels like she's there to ensure the design conforms to the company style guide, and the engineers feel like they are there just to code, and you're back to waterfall.

Finally, collaboration is also *not* about compromise. If you end up with a mediocre user experience, slow performance and limited scalability, and dubious value for customers, you lose as a team.

You need to find a solution that *works*. By that we mean that the solution is *valuable* (valuable enough that target customers will actually buy it or choose to use it), is *usable* (so users can actually experience that value), is *feasible* (so you can actually deliver that value) and is *viable* for your business (so the rest of your company can effectively market, sell, and support the solution).

Accomplishing all this requires real collaboration.

Remember that your job in product is to solve the problems you are asked to solve, in ways that your customers love yet that work for your business. That's your job as a cross-functional product team, and each member of the team is there because they bring specific and necessary skills.

This all starts with true and intense collaboration between the product manager, the product designer, and the engineers.

Our favorite way to do this is to sit around a prototype (usually created by the product designer) as a team to consider and discuss the proposed solution on the table: The designer can consider different approaches to the experience; the engineers can consider implications of different approaches and the potential of different enabling technologies; and the product manager can consider the impacts and consequences of each potential direction (e.g., would there be privacy violations, or would this be something that would work with your sales channel?).

Notice that none of the people on the product team are there to tell others how to do their jobs. Rather, in a healthy and competent team, each member of the team is counting on the others to bring the necessary skills to the table.

But please don't misunderstand.

Designers often have insights based on *deep understanding of your users and their behaviors* that lead you in a different direction in terms of the problem you're solving or your approach to the problem. These insights often will have big impact on value, and indirect impacts on things like performance.

Similarly, strong engineers have *deep insights into the enabling technology* that often lead you to entirely different solutions to the problems you were assigned, often much better than anything the product manager, the designer, or especially the customer could have imagined.

And likewise, the product manager has *deep understanding of the customers and the business* that may lead you to entirely different solutions.

If we had to pick the one thing we love the most about the feeling of true collaboration on an empowered product team, it is the magic that happens when you have people who are motivated, and skilled in their respective discipline—product, design, and engineering—and they sit around a prototype or watch a user interact with the prototype, and the engineer points out new possibilities, and the designer points out different potential experiences, and the product manager weighs in with the sales- or financial- or privacy-related implications, and after exploring several approaches, together they find one method that truly solves for all concerned.

Collaboration means product managers, product designers, and engineers working together to come up with a solution that solves for all your constraints—this is what we mean by solutions that our customers love, yet work for our business.

Getting good at true collaboration is at the heart of how strong product teams work.

CHAPTER

16

Product Strategy

If you have an inspiring and meaningful product vision that you're all working toward achieving over the next several years, the product strategy is your path to making this product vision a reality.

If product teams are all about solving hard problems, then product strategy is all about how you decide which problems are most important to solve.

There are always many good opportunities you could pursue, and there are also many legitimate threats to your business.

The question is: How do you select the best opportunities, and how do you decide which are the most serious threats?

Principle: Focus

As Steve Jobs said, *"People think focus means saying yes to the thing you've got to focus on. But that's not what it means at all. It means saying no to the hundred other good ideas that there are. You have to pick carefully. I'm actually as proud of the things we haven't done as the things I have done. Innovation is saying no to 1,000 things."*

In stakeholder-driven models, it is inherently nearly impossible to have this necessary focus. That's because each stakeholder has their own goals and needs, and the company is simply trying to satisfy as many stakeholders as possible.

In contrast, in the product model, you need to look holistically at both the opportunities and threats, and the product strategy is where you get serious about focus and making the biggest impact.

It's important that the senior leaders of the company participate in the decisions on focus, and it's especially important that the product leaders are not perceived as having an agenda here. This is why it's so important for the product leaders to be transparent.

It's also important to realize that it's often less about which particular goal is declared to be the most important, and more about that a declaration has been made at all. In most companies, there really are several excellent goals, and accomplishing even one of them can improve the company dramatically. But chasing too many of them very often results in none of them being successful.

We often coach product leaders to encourage the CEO to select the two or three most important goals to pursue, and then the product leaders can focus on executing on those choices.

The Power of an Inspiring Product Vision

One very helpful technique for helping with focus is to create a product vision. The product vision describes the common, shared future you are trying to create.

No matter how many product teams you might have, you want each team to know how they contribute to the greater whole.

The product vision typically tries to describe a future between three and ten years out.

Most important, the product vision describes how the world will be better from the point of view of *your customers*. The product strategy is from the point of view of your company, but the product vision is all about your customers.

(continued)

(continued)

There are many significant benefits of a well-crafted product vision. It is very inspiring to the product organization, and often is your single best recruiting tool for new product talent. You can learn more about creating a strong, inspiring product vision in the book *EMPOWERED*.

We mention the product vision here because it is very helpful for focus and prioritization. If the work or product idea in question does not take the company closer toward realizing its product vision, then you would question why this is a priority.

There may still be reasons for pursuing it, but you would want to be conscious that this is a detour on the road toward the product vision.

Principle: Powered by Insights

While it takes real *discipline* to focus, it takes real *skill* to identify the key insights that will power your product strategy.

Insights provide the leverage points that show you how to concentrate your efforts. While these insights can come from anywhere, there are a few main sources:

- *Analyzing the data.* Data about how your customers are using your products. Data about how your customers are buying your products. Data about how this is changing over time.

- *Talking to your customers.* Not asking them what they want you to build (they don't know what is possible) but asking them about what solutions they use today, about their environment and context, and what it would take to get them to switch to a new solution.

- *New enabling technologies.* What can your engineers solve today that you couldn't solve before? What new opportunities do the new technologies open up? What new experiences do the new technologies enable?

- *The broader industry.* What can you learn from the broader competitive landscape? What trends are impacting your industry or related industries? How are customer expectations changing over time?

The product leaders are immersed in these insights, and the entire company should be encouraged to share any insights they might come across with these product leaders immediately.

The product leaders are responsible for aggregating and analyzing these insights, but the insights can and do come from anywhere.

Principle: Transparency

Remember that in the product model, deciding which problems to solve is effectively moved from being distributed across the various stakeholders to the product leaders. These product leaders look holistically at the business and work with the executives and stakeholders to come up with the most impactful product vision and product strategy, in order to pursue the most valuable opportunities and the most serious threats.

It's easy to see how this might cause jealousy or frustration if the stakeholders worry that the product leaders are pursuing their own interests and agendas. This is why it's so important for the product leaders to be open and transparent on the data and reasoning they are using to make these product strategy decisions.

It also should be clear that the product leaders will need to work hard to gain buy-in for their product strategy from the different impacted stakeholders across the company, especially since executing on any product strategy typically entails cooperation from across the company. (You'll see multiple examples of this in the case studies in this book.)

Remember that a product strategy will be effective only if you can execute that strategy.

There is no formula for an effective product strategy. Such a strategy requires immersing in the insights, the data, the customers, the technology, and the industry trends and learnings; assimilating as much of the relevant information as possible; and then thinking through the various options.

This is often referred to as *product sense*, but it's really the result of spending all this time immersed in the details.

The benefit of this analysis and transparency is that the rationale for the choices is clear, and the organization is aligned, all of which enable the organization to execute on this product strategy much more quickly and effectively.

Principle: Placing Bets

Finally, it's important for everyone to realize that product strategy is not a science.

Even with the best product strategy, and with very skilled product teams and product leaders, you should expect that not every product team will solve every product problem like clockwork every quarter.

Realize that every product team is composed of people with different skill sets, working on different problems, with different technologies, and especially with access to different data.

Further, some problems are simply much harder than others. Some problems have a much different risk profile than others. And some problems have much tougher goals to achieve.

The mistake made in the IT model is to ignore this reality, and then act surprised when projects end up taking far longer than expected or fail to deliver any meaningful results.

In the product model, you accept this as a reality of technology-powered products, and you plan for these eventualities.

One useful metaphor for dealing with this reality is the notion of placing multiple bets. In other words, if you have a critically important problem to solve, you might assign that problem to multiple product teams in the hope that at least one of them will make substantial progress in the quarter.

If you get lucky, and multiple teams make substantial progress—sometimes even more than you hoped—that would be an exceptionally good outcome, but it is not a likely outcome.

Experienced product leaders will manage the portfolio of risks by placing a series of bets each quarter, working to maximize the likelihood of meeting the company's annual business objectives by the end of the year.

17

Product Discovery

Product teams are responsible for both figuring out the best solution to the problems they've been asked to solve, and then building and delivering those solutions.

The former is *product discovery*, and the latter is *product delivery*.

Principle: Minimize Waste

The first principle in product discovery is to try to solve the problem with a minimum amount of wasted time and effort. This is why product discovery is the key to accelerated time to money.

The product team could certainly just take their best guess at a solution, describe that solution to the engineers, and then have them build and deliver that solution.

This is essentially what has been done for decades by so many companies in the IT model.

But we also have decades of data clearly showing that the majority of solutions built that way (generally 70 to 90 percent) end up *not* delivering the necessary business results.

This is wasteful not only from a direct cost perspective (especially because the largest cost is usually the engineers), but also from an opportunity cost perspective.

Every once in a while, the team gets to take a second shot, but even then, the odds are against success with that way of working.

The concept of product discovery emerged because smart companies saw the waste in working this way. They wanted to ensure that they had *sufficient evidence* to believe that the solution they asked their engineers to build would successfully solve the customer or business problem it was intended to solve.

The key idea behind product discovery is to test out product ideas very quickly to identify a solution worth building, and then to get that solution to market. The intention is both much faster time to money and much better business results, through dramatically less waste.

Principle: Assess Product Risks

There are always risks in building products:

- *Value risk*—You might build the product only to find that your customers don't feel it's sufficiently better than what they currently have, and they're not willing to buy it or choose to use it and switch.

- *Usability risk*—You might build the product only to find that your users can't figure out how to do what they need to, or are otherwise not aligned to how they think about the problem because the product is too confusing, or not behaving as they would expect, or the learning curve is too steep.

- *Viability risk*—You might build the product only to find that there are legal, compliance, partnership, or ethical issues you hadn't realized that prevent you from selling this product. You might build the product only to find that your marketing and sales organizations aren't aligned with what this product needs to get into the hands of customers. You might build the product only to find that operational costs are too high and/or you can't generate sufficient revenue to sustain this business.

- *Feasibility risk*—You might find it takes so much longer to build than you had originally thought that your costs become so high you can't afford to actually build and deliver this product.

Some problems you're working on have very minor risks, and others have several very major risks. But every effort has risks you need to consider.

Most important, the key principle here is that you need to assess and address these risks *before you decide to build anything.*

Assessing Ethical Risk

Ethical risk is part of business viability risk, along with other important risks such as go-to-market risk and monetization risk. But unlike the other risks, ethical risk rarely has a dedicated stakeholder in the company. Occasionally you will find a Chief Ethics Officer, but today this is the exception rather than the rule.

You might be wondering why the product team should concern itself with ethical risk. The reason is because often the product team is the first to realize during product discovery that there may be environmental, social, or safety-related consequences of the work.

Most of the time, these consequences are not in any way intentional, but nevertheless they may cause externalities or other effects. If you can identify and predict these effects, you can seek alternative approaches without these problematic side effects.

In these cases, often product leaders must get personally involved to help resolve the issue. It is also common for product leaders to coach their people on how to identify and work through ethics issues.

Principle: Embrace Rapid Experimentation

Taking these product risks seriously requires that you're honest with yourself about what you can't possibly know without some investigation, and also what things your customers and your stakeholders can't know (usually because they don't know what's technically possible).

Without creating a prototype and running some experiments, it's very difficult to know if customers would really be able to learn and use a new product, or whether they actually would buy this new product.

Without the engineers having some time to investigate the underlying technology and the potential solution, it's very difficult for them to know how long something might really take to build.

The heart of product discovery is rapidly testing product ideas for what the solution could be.

You *think* you understand the solution that will solve your customer's problem, but you are aware and humble enough to know it's just a guess. So, you create an experiment designed to quickly determine if this approach will or will not work.

An experimentation culture not only helps you address risks, but it is absolutely central to innovation. Often the idea that succeeds started out looking like an idea that had little chance.

A culture of rapid experimentation encourages product teams to pursue these ideas. Skills in experimentation enable the product teams to test these ideas quickly and inexpensively.

More generally, in the product model, it's normal for people to disagree.

In fact, if members of the product team and the stakeholders they collaborate with actually care about what they're doing, disagreements are a good sign.

But that also means there will be lots of disagreements. For that reason, product teams need to be skilled in the many discovery techniques used to run quick tests to collect some data so that you can make an informed decision. And they must develop the judgment to know how much data they need to collect for the particular situation.

This means they need to be skilled at both quantitative and qualitative discovery techniques. The quantitative techniques let you learn *how* your users and customers actually engage with your product ideas, which is a tremendous asset. However, the big limitation is that this data generally can't tell you *why* customers are using (or, very often, *not* using) your product ideas. For this reason, you also need qualitative techniques, the most important of which is actually talking to your users and customers.

A corollary of this is that everything you build and deploy needs to be instrumented such that you know how your products actually are being used. Without this data, you are flying blind.

With this data, you can improve the product even faster, and also quickly diagnose issues that may arise.

Principle: Test Ideas Responsibly

Product discovery techniques are used in companies of every size, from the smallest startups to the largest enterprises. But there are some differences. Some things are easier at a startup and other things are harder.

Startups often don't have much traffic, which hinders their ability to collect a sufficient volume of data.

On the other hand, large, established companies usually have many customers. Assuming the company has instrumented its products and is collecting this data, this can be an advantage. However, as discussed earlier, if the company is not yet collecting this data, this is at the foundation of *changing how you build*.

Another difference is that startups usually don't have much revenue and don't have many customers. Until they correct that, there's very little for them to lose. But in a larger enterprise, you have a great deal to lose.

That is why one of the principles of product discovery in established companies is the need to assess the risks, run your experiments, collect your data, and iterate quickly—all while ensuring you are protecting:

- The company's revenue
- The company's reputation
- The company's customers (from being confused or frustrated)
- Your colleagues (e.g., sales or customer success) (from being blindsided)

For techniques where any of these are a concern, variations exist that you can use when you need to be more conservative in your testing. To be clear, you absolutely need to run the experiments, but you need to do so responsibly.

18

Product Delivery

Most companies moving to the product model not only have to introduce the tech lead competency we discussed earlier, but they also have to upgrade their skills and infrastructure when it comes to building, testing, and deploying the products that their customers depend on.

In strong product companies, you'll often hear the phrase *"Reliability is our most important feature."*

In modern technology-powered products and services, breaking the product can have immediate and damaging consequences to your users and customers, your revenue, your brand's reputation, and your colleagues (especially sales and customer success staff).

If you cause a serious issue, it can create an outage for all the customers and users who use and depend on that service. This is part of the price we pay for the many benefits of cloud computing.

With every product, there will be times where your customers encounter some serious problem requiring immediate action.

In these cases, you need the ability to immediately roll back the last change to stabilize the system, quickly diagnose the issue, create a solution, test that solution to ensure it solves the problem and doesn't inadvertently cause other issues (regressions), and then safely deploy that solution.

Customers generally understand that occasionally problems will occur, but they judge you more by your ability to respond quickly and competently when those issues do arise.

Waiting weeks or months is simply no longer acceptable for most companies today. Strong product companies need the ability to respond quickly and competently to pressing customer or market needs.

To serve your customers, you need to ensure your products are working properly and generating the necessary value. This is what drives the need for a set of critical capabilities to support deployment, instrumentation, monitoring, and analytics.

While the concepts and principles in this chapter are somewhat technical in nature, they are all essential and, at the conceptual level, fairly straightforward to understand.

Principle: Small, Frequent, Uncoupled Releases

The core principle that is used to address these needs is the deployment of *small, frequent, uncoupled releases.*

This means, *at a minimum*, each product team releases their new work no less than every other week. For strong product companies, this means releasing *several times per day* (referred to as CI/CD—continuous integration, continuous deployment).[1]

The reason you likely don't know this is happening with your favorite products is precisely *because* they are releasing a near-constant stream of very small releases.

But ensuring reliability is somewhat more complicated than it may seem. In general, you work to test two main aspects of what you build—the first is straightforward and the second is less so.

---

[1]The most common objection to this comes from product teams building native mobile applications, where they point out that app stores don't like when you submit new versions of your app more than once a month or so. But we have tools for several years that allow us to release continuously to a controlled set of customer devices so that we get the benefits of continuous deployment.

The first is that when you build a new capability, you need to test that the new capability will work as expected. Straightforward, although since you likely will want to test and retest this new capability thousands of times in the coming months and years, you will usually invest in some level of test automation.

The second is to ensure that the changes made to enable the new capability do not unintentionally or inadvertently break anything else. This is referred to as *regression testing*. And when you realize that many technology-powered products and services today are the result of literally hundreds of engineers working for many years, creating tens of thousands of interactions, you can see that ensuring a complex product does not take a step backward when a new capability is introduced can be a significant undertaking.

The main method product teams use to ensure that new capabilities work as advertised, and don't introduce regressions, is to deploy a series of very small changes.

The smaller the release increment, the faster you can ensure the quality of the new capability, and the faster you can be confident you're not introducing regressions.

And with these small, frequent releases, if a problem is introduced, it is much easier to identify the cause (since you only changed a small number of things).

So, if you are serious about taking care of your customers, you will need to invest in your ability to deliver very frequent, very small releases.

If this seems counterintuitive to you, and if you still believe that to ensure quality you need to release slowly and infrequently, then you owe it to yourself, your company, and especially your customers to examine the theory and the evidence on why frequent, smaller releases provide both more throughput (higher velocity) *and* higher quality at strong product companies.[2]

---

[2]We recommend the excellent book *Accelerate: The Science of Lean Software and DevOps: Building and Scaling High Performing Technology Organizations* by Nicole Forsgren, Jez Humble, and Gene Kim (IT Revolution Press, 2018).

Unfortunately, many companies have not yet achieved the ability to do these small, frequent releases.

Instead, they make hundreds or even thousands of changes, and then, once a month, once a quarter, or even once a year, they will work to integrate all these changes together. They then try to test that all these new capabilities work as expected, and then they begin the grind of trying to identify and remove all the newly introduced problems (regressions).

As we hope you are beginning to realize, this is why large releases (known as *big-bang releases*) are notorious for delays of weeks or even months, trying to get everything back to a reliable, releasable state.

In fact, many products built this way *never* achieve a state of solid quality, and the customer is forced to deal with a constant stream of defects and issues, or else seek out a different solution provider.

Moreover, even if the new release actually works as it's supposed to (a very big *if*), customers are forced to absorb hundreds or even thousands of product changes hitting them all at once. This likely requires retraining, recertification, reintegration, and otherwise significantly interrupts their own work to accommodate all the changes the company has forced upon them.

In this case, it's common for customers to press your company to release *less frequently*, because they simply have no time to deal with this degree of change. As you can see by now, while it is completely understandable why customers would request fewer releases, doing so actually would be *worse* for the customer and *worse* for your company.

However, rather than ignore or discount customers' legitimate concerns, the responsibility is on you to design, test, and release your work so that customers do not have to deal with the impact of the changes. And while you have good reason to get your changes into the production environment as quickly as possible, you also have techniques to control when customers can see and access these new capabilities, which we'll describe in the principles that follow.

High-Integrity Commitments

In the product model, product teams are all about solving hard problems for your customers and your company, in ways that your customers love, yet work for the business.

That said, the product model also realizes that sometimes you need to deliver a specific deliverable on a specific date. These are called *high-integrity commitments.*

Perhaps there's a partnership involved, and the partner must be able to plan their work from their side. Or perhaps your company is planning a major marketing campaign or industry event, and so you need to commit to being ready.

It is no secret that typical Agile teams have a terrible track record in terms of delivering on these types of promises.

But strong product companies know how important this is to building and maintaining trust with customers, partners, and the rest of your company.

If your customers or partners are depending on you for a specific deliverable, they need to know they can count on you to deliver when promised, and that the deliverable will provide the necessary value.

In this case, you want both to solve the underlying problem and to determine with high confidence when you can commit to a deliverable on a specific date.

Here's the key: So long as the product team responsible for making the high-integrity commitments is first allowed to do sufficient product discovery, such that they can reasonably address the product risks, then they can safely and responsibly commit to a deliverable on a date.

If the engineers are experienced or have competent coaching, this almost always means the product team will first create a *feasibility prototype* before they commit to anything. This popular technique is described in the book *INSPIRED.*

Further, it's absolutely essential that *the people who provide the dates are the ones who will actually have to deliver the product.* These dates should *never* come from project or program managers, product managers, architects, or anyone other than the people who will have to do the actual work.

Moreover, since the CTO (head of engineering) is ultimately responsible for the commitments made by her engineering organization, best practice is that she should personally approve all high-integrity commitments.

It's worth emphasizing that high-integrity commitments, when used appropriately and judiciously by the company's senior leadership[3] and when handled professionally by the product team, can go a long way in building the necessary trust across the organization.

Principle: Instrumentation

Because in the product model you are committing to solving problems and are held accountable to achieving outcomes, it is essential that you understand how (or even if) your products actually are used.

This means that you need to ensure your products are *instrumented* such that you are able to understand what is going on. This is referred to as *telemetry*, and it happens at all levels—from low-level services reporting on their health and performance, to high-level applications generating usage analytics and company dashboards.

Without this instrumentation, you are simply flying blind.

You can release a new capability and have little idea if and how it is being used, and where your customers might be struggling. With good instrumentation and analytics, you can quickly detect and correct problems, and demonstrate that you are providing the necessary value.

Many different tools and services are available to your engineers to provide the necessary level of instrumentation. It is normal to need multiple tools because of the different types and levels of

---

[3]It's also important to acknowledge that some companies try to make everything a high-integrity commitment, which completely misses the point, and actually undermines the product model.

instrumentation, but it is critical to understand the importance of this telemetry data.

It is also important to recognize that the specific telemetry data you collect is always evolving as you get a deeper understanding of how your products are used, and the improvements you need to drive. This is something that is never "done," but rather is constantly improving.

Principle: Monitoring

There are many benefits to the instrumentation described in the preceding principle, but one of the most important is that instrumentation enables another delivery principle: *monitoring*, also known as *observability*.

As with instrumentation, monitoring happens at all levels—from ensuring that the underlying computing systems and services are operating and performing properly, all the way up to ensuring that your application is behaving correctly and serving customers appropriately.

With strong monitoring, you can very quickly detect issues, often before your customers ever encounter a problem.

As with instrumentation, a wide range of commercial tools are available to help with monitoring. And as with instrumentation, you can expect your company to need multiple monitoring and reporting tools for the different types and levels of information.

For both instrumentation and monitoring, the tools use techniques to prevent tracking or reporting any sensitive or personally identifiable information.

Principle: Deployment Infrastructure

So far, you have the ability to deploy small, frequent releases that are instrumented to provide the necessary telemetry data, and you are monitoring the status of these releases.

But to ensure you are providing the necessary value, there is one more important element, which involves the infrastructure used for deployment.

Suppose you have a new capability that is ready to be released into production. Even assuming you have tested that new capability and ensured that it works properly, occasionally you will find that in the production environment, there is some sort of problem, and the deployment infrastructure needs the ability to roll back that change when required.

But even if things are operating fine, what you don't yet know is how this actually will be utilized by your customers in their daily use.

There are three possible outcomes here:

1. Customers love this new capability, and they immediately start using and depending on it. This, of course, is what you hope for.

2. Once it's released into production, this new capability for some reason actually *hurts* your customers' ability to use your products. One common reason for this, especially with mobile applications, is that due to the very limited amount of real estate on their device, sometimes when you add a capability, customers can no longer find a more important capability. In this case, you would not want to deploy this capability until and unless you could correct this problem.

3. The new feature is deployed and technically works fine, but it's simply not used, at least not to the level you had hoped for or expected. The result is that the new capability does not hurt or help anything. This case is actually surprisingly common and very frustrating.

If you are to hold yourself to a standard of achieving business outcomes, then you need to deploy new capabilities such that you know which one of these three cases is happening.

The most common way of doing this is to have your deployment infrastructure support A/B testing. This is considered the "gold standard" because you can easily isolate the contribution of just a single new capability. However, this also depends on significant levels of traffic to determine results quickly.

You may not be aware of this, but companies operating in the product model are running many (often hundreds or more) of these tests *simultaneously*. The deployment infrastructure runs these various tests, collects the analytics for each test, and continues until there are statistically significant results.

Other deployment infrastructure capabilities enable you to control which of your customers are able to see specific new capabilities. One especially popular capability allows you to deploy new capabilities into production, but to keep that new capability hidden from your customers until such time as you are ready. This technique is especially helpful when you have a combination of changes that need to be made visible simultaneously or when you want to wait to show it at a special marketing event.

In terms of deployment infrastructure, although several commercial tools are available, companies have a range of needs, and it is not unusual for companies to have a blend of commercial and custom infrastructure.

Managing Technical Debt

We mention the tech debt problem here because if the situation is severe enough, this will get in the way of everything else, starting with changing how you build and deploy.

Every company has technical debt,[4] but a disproportionate number of companies that want to transform have very serious level of tech debt.

There are two common reasons for this:

1. Companies that have a history of acquisitions not only inherit the debt of the companies they acquire, but between the need to integrate with other systems and the postacquisition loss of experienced engineers, the problems usually multiply.

---

[4]Likely you have heard of the concept of *technical debt*, but it is the cumulative impact of years' worth of both growth as well as engineering and architectural shortcuts in the name of expediency. To be clear, this is normal and, to a degree, even desirable, but if left unaddressed, tech debt can literally destroy companies. The most common symptom of this problem getting out of hand is when work that would normally take days starts taking weeks, and you hear the frustration this causes across your company and your customers.

2. In most of the prior models, especially those run under the classic IT-style project-based funding model, this is a recipe for creating very high levels of technical debt quickly, as funding is only for the particular project, and usually does not include looking holistically at the systems to constantly improve products and the systems that run them.

No matter the cause, the symptoms of tech debt are usually quite clear. Work that used to take days or weeks now takes months. Product teams complain of too many dependencies and too little autonomy. Even small capabilities become expensive, and the large initiatives that are so important to companies become so difficult that the company tries everything possible to avoid them.

This is why technical debt is so feared across so many companies. It is one of the few true business-continuity risks that companies face.

Moving to the product model where product teams have real ownership of major areas is a big step toward improving the situation, but you still will need to dig yourself out of the hole.

The specifics of different replatforming efforts are well beyond the scope of this book, but specialty firms can help you formulate a solid plan to modernize your code base, if needed. However, the most important thing to realize is that you need to start on this work on addressing tech debt immediately, and then do this work continuously, indefinitely.

Most companies with a good handle on tech debt will tell you that they work on tech debt every day, with about 10 to 30 percent of the engineering capacity. Sometimes, the situation is so dire that it demands more like 40 to 60 percent of the engineering capacity. At this point, you have very little remaining capacity for your customers, so you would need to proceed carefully and intentionally.

While getting this tech debt issue under control usually will take one to three years, the good news is that if you have a solid plan and experienced engineers, you often will begin to see real results in just a few months.

CHAPTER
19

Product Culture

W e've been discussing the product core concepts of product teams, product strategy, product discovery, and product delivery, but if you consider all the core principles behind each of those concepts, you can get a good sense of what is meant by a strong product culture at a strong product company:

Product teams empowered to figure out good solutions to hard problems, staffed with the cross-functional set of skills necessary and provided unencumbered access to customers, data, and stakeholders.

Product strategy powered by insights that help the organization decide the most valuable opportunities to pursue and the most serious threats to counter.

Product teams skilled at *product discovery* by assessing risks up front, embracing experimentation, prototyping, and rapid testing with users and customers to quickly determine which solutions actually are worth building.

And once product teams decide to build, they have the *product delivery skills* to build, test, and deploy via small, frequent, reliable releases that are instrumented and monitored.

There are, however, a few other critically important product model principles can be thought of as meta-principles, as they apply to product development more generally yet play a very important role in defining a strong product culture.

Principle: Principles over Process

Many companies that want to transform were, in their earlier years, quite good at innovation, but then somehow, they lost this ability.

Unfortunately, this is all too common, and it is something that strong product companies are always fearful will happen to them.

As Jeff Bezos warns: *"Good process serves you so you can serve customers. But if you're not watchful, the process can become the thing. This can happen very easily in large organizations...The process is not the thing. It's always worth asking, do we own the process or does the process own us?"*

Or as Steve Jobs warns: *"That's what makes great products. It's not process; it's content...The system is that there is no system. That doesn't mean we don't have process...But that's not what it's about."*

Or as Reed Hastings warns: *"[The reason Netflix has been so successful is because it has] a culture that values people over process, emphasizes innovation over efficiency, and has very few controls."*

Or as Steve Blank warns: *"Process is great when you live in a world where both the problem and solution are known. Process helps ensure that you can deliver solutions that scale without breaking other parts of the organization...These processes reduce risk to an overall organization, but each layer of process reduces the ability to be agile and lean and—most important—be responsive to new opportunities and threats."*

It's not that process is inherently bad, but many leaders and managers will join your company from companies that long ago lost their ability to innovate, and they bring with them the processes and culture that they came from. Or, they bring processes from the nontechnology space and try to impose them on organizations trying to optimize for innovation.

Fundamentally, you can try to run your product organization with process or with people.

If a person makes a mistake, you can either add a process that is intended to prevent that mistake from happening again, or you can provide some coaching so the person understands how they might better handle the situation going forward.

If you're trying to empower your people with a true sense of ownership and push decisions down to the product teams and closer to the users, then you'll need to make coaching a pillar of what your managers are responsible for.

Coaching includes teaching skills and principles, and also sharing the strategic context—the larger picture of the product vision and product strategy.

Continuous Process Improvement

Another way of ensuring that the principles we're discussing here are always more important than any particular process you follow is to intentionally practice *continuous process improvement*.

The idea is to always reflect on your experiences and your needs, and constantly strive to get better. This means being careful you don't fall into the trap of becoming religious about process.

Principle: Trust over Control

Moving from the command-and-control model to the product model is not only a change in competencies and concepts, but it represents a fundamental cultural change.

For many company leaders, especially those who built their careers on the top-down, command-and-control model, this is a big leap. It is a model that depends on trust rather than control.

It can be helpful to realize that this is what's really going on underneath this very different way of working.

This approach shows up in many ways, starting with moving from stakeholders providing product teams with prioritized lists of features and projects, to instead empowering the product teams by giving them problems to solve and allowing them to take responsibility for coming up with the best solutions to those problems.

More generally, this means moving from hands-on micromanagement to servant-based leadership with active coaching. It means leading with context rather than control.

Principle: Innovation over Predictability

In many companies, the root cause of their lack of innovation is quite plain to see. They have designed their organization and how they work around the goal of predictability. They simply focus on being able to ship large numbers of features every quarter.

It's not that they are intentionally abandoning the goal of innovation, but as Henrik Kniberg says, *"100% predictability = 0% innovation."*

Companies that focus on predictability will tell you that it's the stakeholders' job to worry about whether the features shipped are valuable or not. But strong product companies know this is not a question of smart stakeholders. It goes to the nature of technology-powered products, and especially that stakeholders and customers don't know what's possible. It is the engineers who know what is just now possible.

Today, more than ever, companies depend on continuous innovation. So, recognize that predictability is good but not as important or necessary as innovation.

The best way to handle the need for occasional predictability of a deliverable is with *high-integrity commitments*.

From Projects to Products

Talking about product culture can sound very theoretical, but it impacts everyone every day. To try to bring these principles to life, here is an example of a very common before and after.

(continued)

In the old models, it is very common that everything is about building *projects* by certain *dates*.

It's understandable where this comes from, as the IT department is constantly asked for specific features and projects delivered as fast as possible.

The problem is that this ignores the realities of technology-powered products.

Projects are typically large, slow, and expensive attempts to deliver some output by a specific date. You have to decide how big the team needs to be, and guess how much time the project will take. And then you need to try to get funding for the effort, and invariably you learn that there is more work required than you first expected, so there's a real crunch to finish.

Most efforts lose all hope of providing real value and just try to get something shipped. Then as soon as they do finally ship, they can't iterate to improve the product because the people usually scatter off to their next assignments. Nobody owns the outcome, and any important learnings from the effort likely are lost.

There are so many things wrong with this:

1. It normally takes several iterations, based on learning from prior iterations, until you achieve the necessary results, but you rarely get funding for more than the first attempt. Even if you do, it's usually multiple quarters later.

2. You need the people to feel real ownership of the technology and the outcomes, but when people are assigned to an area just for the duration of the project, there is very little ownership, and there's little incentive for them to build something impactful.

3. It's all about delivering the features and projects that the stakeholders are asking for, rather than allowing your product team, especially the engineers, to look at what's just now possible. The result is that any form of innovation is exceptionally rare with projects.

4. Because teams build just what they need to for that project, there's no one to worry about the longer-term impact of this work,

(continued)

or improving the underlying technology, so tech debt rapidly accumulates.

5. Who takes responsibility for driving to the necessary outcome? Any sense of ownership usually is dispersed among multiple stakeholders, and the most likely consequence is to blame the project team for doing such a poor job.

In contrast, with the product model, you focus on *products* and *outcomes*.

A product team focuses on business outcomes, like reducing churn rates, or improving growth, or whatever the relevant KPIs are for your situation, and the product team is constantly working to monitor and improve those results. The capabilities they add are all intended to drive those results. They are obsessing over the outcome.

In part, this is about changing to durable product teams focused on business results, but, in part, this is a cultural change where a team is asked to be accountable to *outcomes* and not just shipping features.

In the project model, the best you can really ask for is time to market. But in the product model, you can focus on the much more impactful *time to money*.

What is especially ironic is you might think that a time-to-market *project* at least would get done *faster* than a time-to-money *product* team would achieve an outcome. But so often it is just the opposite.

While the project team must staff up, learn the necessary technology, establish the necessary relationships, and understand the necessary context, a product team is already operational and likely has done many similar projects and features already. So, in addition to being up to speed on the technology, they know much more about both the problem and solution space, they know the data, and they know how to work as a team to tackle a problem.

This is why, far and away, the most waste we see is in companies running in the old project model.

None of this is to say that time to market isn't important. It is. And speed matters. And there are solid techniques when a date you can count on is the priority.

(continued)

That said, the heart of the matter is the question: What is more important for this effort: hitting this date or accomplishing this outcome? Sometimes the date *is* the priority. But this should be the exception and not the rule.

It's easy to see why so many companies may talk about the importance of outcomes over output, yet their culture and behaviors consistently prioritize predictability over results.

Principle: Learning over Failure

In many companies, there is a deeply rooted fear of failure. This fear drives people and processes to become risk averse, and unable to respond to the changing needs of the market and the new enabling technologies.

While it's never our intention to glamorize failure, the focus must be on tackling risks and learning quickly. When you run an experiment in product discovery, there is no such concept as *success* or *failure*; there is only the question "What have we learned?"

Your goal is to learn what works and what doesn't in product discovery, where costs and time are dramatically reduced, and risk is mitigated. This is so when you do move forward with the time and expense of actually building products, you have evidence and confidence the product won't be a failure.

Yet it's also true that with some product efforts, you still may fail. A strong product culture understands that risk, and embraces the learning and the people who were willing to take the risk.

20

Innovation Story: Carmax

Marty's Note: CarMax had already been seeing the benefits of their transformation to the product model when the pandemic put the organization to a true test. It required the company to use all its new muscles—product strategy, product discovery, and product delivery. Sometimes it takes a crisis to see what an organization is truly capable of.

Company Background

CarMax, founded in 1993 and based in Richmond, Virginia, is the largest retailer of used cars in the United States. The company has built a strong business over the past 30 years by bringing trust and integrity to buying and selling used cars.

CarMax was initially created as a spin-off of electronics retailer Circuit City and quickly grew to become a leader in used car sales. Over time, the company's leaders saw consumer expectations beginning to change, especially around the desire for an online shopping and buying experience, as well as the emergence of digital-native competitors, and they realized that they would need to disrupt themselves to maintain their leadership position.

The company's initial efforts in moving to the product model were focused on the consumer-facing digital experience, which was led by Ann Yauger. Ann instigated and led the initial transformation to the product model with the carmax.com team. The initiative quickly proved its value—starting several years of organizational growth to build out product competencies at scale.

By the start of 2020, the company had more than 200 CarMax stores, each with hundreds of used cars for sale on its lots. The company was consistently profitable and growing.

The transformation had already produced tangible results, and CarMax built a reputation for innovation in the retail car space.

Consumers could research available cars online, discover what cars they could afford, and make an appointment to visit a store to test drive specific cars.

Unlike most car purchase experiences, at CarMax, there was a simple, no-haggle price for each car they would buy or sell, and each car CarMax chose to recondition for retail had to pass an extensive 125-point inspection, with high standards for quality.

Cars that made the grade were available for sale to CarMax customers, along with a 7-day money-back guarantee and a 30-day warranty.

Cars that did not pass this level of quality were sold via auctions to other dealers.

The Problem to Solve

When the pandemic hit, many brick-and-mortar retailers—including CarMax—needed to comply with a variety of state and local restrictions and mandates. This included in some cases closing stores, reducing the number of customers allowed in the store at the same time, or only selling vehicles outside.

With these new restrictions and the broad macroeconomic impact from the pandemic, the company shared that by early April 2020, sales were down more than 75 percent versus the previous year.

Yet, in addition to the challenges other physical retailers faced, CarMax held billions of dollars' worth of inventory—cars sitting on lots—but with real limitations on how the cars could be sold.

Fortunately, CarMax had already begun to move to an omnichannel retail experience, where customers could purchase either online or in a store, or via their desired blend. But now this was no longer a business growth opportunity; it instead became business critical.

The company needed to accelerate its move to omnichannel retail and discover additional ways to sell its tremendous inventory of cars within the constraints of the pandemic.

Discovering the Solution

At this point, CarMax had several strong product teams, many that had already been working on areas related to the omnichannel capabilities that were now the priority.

The most obvious changes had to do with moving to an experience that was not dependent on visiting retail stores and interacting live with in-store sales associates.

The company accelerated its move to a virtual customer experience center, staffed with associates who could provide assistance at any stage of the car-buying or car-selling process.

The company also needed to offer a way not only to browse cars online, but also to select a car, finance or pay for it, complete the sales order, and arrange delivery or pickup. And if a trade-in was involved, an online way to get an actionable and immediate trade-in value for that car was needed.

Further, CarMax could no longer rely on in-person auctions with other dealers to sell the cars it acquired on trade-in that did not meet its quality standards.

These new car-buying and car-selling experiences comprised several related product efforts having to do with all stages of the transaction. These stages ranged from financing, to making an "instant-offer" for a trade-in, to electronically making the down payment and agreeing to the financing terms via electronic document signature, to arranging delivery logistics including curbside, touchless delivery, or home delivery, and to the new "love your car guarantee," which extended the warranty period from 30 to 90 days, and the return window from 7 to 30 days, which was unmatched in the automotive industry.

All these aspects of the transaction are by definition risky, as they involve a major consumer purchase, but some of the areas had especially high risk.

For CarMax, which has a business model predicated on consistent margins, the company had to ensure that the pricing both for the cars it purchased, and those it sold, was accurate and fair.

CarMax had to be sure it could describe and communicate clearly and accurately the true state of the car, because if the retail customer, or the wholesale dealer buying at auction, was surprised once they saw the physical car, they would lose trust in CarMax and likely not complete the transaction.

To address these risks, the product teams tested prototypes first in very small groups, then regionally, and finally nationally.

Key components of the new sales model continued to roll out as the testing progressed, and within only six months from the onset of the pandemic, CarMax had all the key elements of the car-buying and car-selling process operational nationally.

The Results

After suffering a severe revenue hit at the beginning of the pandemic, CarMax quickly recovered the lost revenue and more—becoming one of the few auto industry innovation success stories.

Over the two years that followed, the new technology-powered services provided a much richer customer and associate experience than the more manual-based solutions used prior to the pandemic.

Today, CarMax has earned a reputation as one of the best places to work for product and technology talent, and the speed and abilities of the product and technology organization helped earn the trust of colleagues across the rest of the company.

None of us can know what the future will bring, but the CarMax team has proven that it can respond quickly and effectively to solve important and difficult problems for their customers and for their company.

Transformation Story: Trainline

By Jon Moore

Marty's Note: In just a few short years, Trainline became one of Europe's best examples of tech-powered product innovation—completely turning around its business in the process. One thing that was a bit unusual is that this transformation started when a large private equity firm, KKR, purchased Trainline because it (correctly) believed the business was significantly undervalued and that if Trainline moved to the product model, the company could unleash real value. KKR brought in a strong CEO, Clare Gilmartin, who then brought in a proven head of engineering, Mark Holt, and a proven head of product, Jon Moore. Here Jon shares a first-person account of this impressive transformation.

In January 2015, one of the world's largest private equity companies, KKR—famed for inventing the leveraged buyout—announced it had purchased Trainline, a long-standing reseller of UK rail tickets.

The acquisition, on the face of it, made limited sense. Rail remains a supremely legacy industry. Ticketing changes can require multibillion-pound infrastructure investments, and the entire industry is subject to close governmental scrutiny. In such an environment, the route to growth is normally measured not in months or years, but in decades.

With these constraints, Trainline appeared to offer little room to achieve the kind of progress demanded by KKR's portfolio.

What followed was a remarkable achievement, one led by a new, young CEO plucked directly from eBay to spearhead a dramatic transformation away from legacy rail and toward a strong consumer-technology business.

The Motivation

When the new CEO, Clare Gilmartin, hired me as the company's first chief product officer, she warned me that every part of the organization would need to change if we were to achieve anything like our expected trajectory.

The company was making little progress. Trainline had a classic IT model doing what it could to serve the business but providing few results.

Software engineering was nascent and, in many areas, outsourced.

Product management was completely absent, save for a few individuals who had been given the vaguely defined title of product owner.

The few designers who did exist described themselves as owning a variety of marketing assets.

The products themselves were woefully outdated.

Originally launched in 1997, Trainline offered a solution to the chaos that followed the privatization of the UK's rail network. Some years earlier, the state-owned British Rail had been disbanded in an attempt to foster innovation and competition.

But it had been rushed, and the resulting network of contracts involved more than 100 companies. The customer entry points were confusing and varied. Purchasing rail tickets often appeared to require an understanding of which company was running the line on which you wanted to travel.

Trainline's centralized online booking platform provided an answer. Allowing customers to skip the daily frustrations of booking office queues, it sold its first online ticket in 1999 and, over the next decade, gradually established a toehold in the market. But then the company stalled.

In 2014, Trainline was purchased by KKR, which saw an opportunity and persuaded a talented modern technology leader to leave eBay and take on her first CEO role. She brought on an experienced chief technology officer, Mark Holt, and me to run product.

The three of us quickly bonded as a team and got to work on our respective areas.

Changing How We Build

Engineering, product management, and design were prioritized, and early executive meetings were regularly dominated by conversations concerning ongoing recruitment needs and talent gaps.

Engineering needed urgent investment. Outsourcing was rife, and the more senior engineering roles were filled not with internal talent, but by multiple contractors who had gained significant knowledge of the various complexities of the platforms and systems, but had little purchase in the future direction of the company.

The organization was operating on eight-week release cycles, not close to being good enough to allow us to serve customers effectively.

We had inherited a classic CIO culture. Technology had simply been treated as a necessary cost.

The consequences were easily visible. Tech debt was rampant; systems were monolithic and fully on-premises. A significant technical turnaround was needed quickly.

Every job spec was rewritten, a growing number of new engineers were recruited, and a new tech lead role was created at the team level. In truth, a significant number of individuals were let go.

The engineering challenge was considerable. Trainline's legacy data center, consisting of hundreds of servers, was unfit for its purpose. Then we received notice that the building we occupied was due to be demolished.

Yet within 18 months, Trainline had been fully migrated to AWS and was accelerating quickly toward being 100 percent cloud native.

In time, over 20,000 individual components were released into the cloud, at a rate of more than 100 a week. "One afternoon in May," Mark remembers, "somebody said, 'By the way, everything is in continuous delivery!' It was a truly fantastic moment."

Engineering was on a fast track, not just structurally but culturally. New recruits were chosen not simply for their technical prowess (our AWS migration had been run by one of the most accomplished cloud experts in the UK) but based on their desire and ability to understand customers.

Engineers were expected to use the product extensively, and to understand as much of the customer and business context as possible.

Changing How We Solve Problems

With impressive engineering talent arriving rapidly, the need to achieve strong product management was critical. I had inherited a team used to delivering on requirements from stakeholders. These servants to the business had little understanding of product discovery or how strong product teams solve problems for customers in ways that customers love yet work for the business.

But the new mandate was clear. We needed a strong PM discipline at the heart of the business. Our intention was to rapidly transform into a machine that could uncover—and then achieve—maximum value.

I had feared we would need a clean sweep of talent. But, on entry, a number of individuals made it clear they wanted to learn how to work like the best. And I quickly realized that one, more recent hire had maximum potential.

As a whole, the team had not been exposed to modern product management, but they were intelligent, ambitious, and anxious to learn. I decided to give them the opportunity to show what they could do.

Coaching became a first-order priority, and multiple hours were spent every day working individually with team members. We discussed the critical importance of product discovery, key techniques, and the mandatory need for a much faster and stronger test-and-learn cadence.

To support the shift in culture, a wider informal meeting was set up every Friday afternoon called "Weekly Wins." My intention was to create a safe space for all teams to come together and discuss and debate progress. It quickly generated its own momentum. Discovery insights, test data, and prototypes were all regularly showcased. Questions were asked and ideas were discussed. Communication and coordination improved significantly. These Weekly Wins rapidly became the meetings nobody wanted to miss.

Everywhere we turned there were multiple issues to be addressed.

Travel is by nature inspirational, yet our products were unquestionably transactional and uninspired.

With so many early potential priorities, we quickly decided that the product strategy should be to focus on two critical core problems: mobile apps usage and site conversion. These areas offered the prospect of immediate and dramatic results.

Trainline was still almost entirely desktop focused, a miss of almost ten years. As a marketplace, conversion expertise was also a clear growth lever, but few tests had been run and there was no data platform to speak of.

The company had been using an external consultancy to spearhead optimization. The monthly cost ran into the tens of thousands of pounds.

When I pointed out that only eight tests had been run in the previous 12 months—at a cost of nearly £40,000 per test, and with none offering any obvious upside—there was little argument to ripping up the contract. We would grow our own muscles in this area, and we needed to do it quickly.

Our newly created mobile apps team led the way, iterating quickly to achieve a product that quickly gained traction within the Apple ecosystem—dominating the category rankings. We were determined to set the bar high, selecting the team carefully to create an internal bar raiser. They did not disappoint.

Led by an impressive new product manager and tech lead, who together brought significant ambition and pace, the break from the past could not have been more visible.

Our mobile usage grew rapidly, and as leadership, we repeatedly celebrated the outcomes and the team.

We realized this might be frustrating for others in the company. But, in offering a concrete example of the new cultural expectations, we were attempting to jolt others into understanding what it meant to focus on business outcomes. And it worked.

With a strong emphasis on recruiting and coaching, we gained momentum quickly. Team by team, others followed their example.

Design as a Superpower

The design team was explicitly recruited to achieve a strong bias toward high-velocity qualitative testing. Prototyping became a key strength. Rapid user prototypes were consistently used across all teams to gain quick feedback on myriad potential solutions.

The team grew quickly, though not without first overcoming some resistance. At one point, finance rejected additional hiring in product design. Understand that this was a private equity–backed business where controlling expenditure is always a focus. But we needed product designers to get the most out of our investment in product management and engineering.

I realized my initial appeal had failed because I had not shared the evidence of the business results. When I shared the improved outcomes we were delivering in the teams with strong product design talent, the design staff hiring was quickly approved.

With so many parallel workstreams, we designed our team topology to lower cognitive load and minimize dependencies. This choice was bolstered by our growing investment in a cloud-based microservices platform, allowing us to minimize—though not eliminate—dependencies.

A variety of experience teams took charge of our main device-based experiences (desktops and mobile), and while this evolved over time with the addition of further teams targeting newer B2B and ancillary growth opportunities, our more detailed conversations were reserved for how we intended to invest in our platform. Specifically, data.

Investing in Data Science

With our data stored across a wide variety of systems in multiple, inconsistent formats, there was no obvious data strategy to build upon.

"Our data is a box of gold with 10 inches of dust on top," I told Clare. "We just need to build the means to access it." I was making the case for yet another significant investment, but Clare already saw the potential.

Despite a long list of competing priorities, Mark forged ahead with the data engineering recruitment, and, in parallel, I took on the challenge of building out our first data science team.

We knew we were investing somewhat ahead of the curve. Beyond the obvious need to be better able to organize our data and supercharge our site optimization workstreams, we had little real concept of what these teams ultimately would be capable of. But we knew enough to understand that, without them, we would be limiting our potential solutions considerably.

Empowered Engineers

To help cement the ongoing transformation toward a modern product culture, we organized our first hack day. Teams were able to focus on any customer or business problem they chose. In truth, the main objective was to encourage the engineers to pursue more of their own ideas and to further enhance the innovation culture we were building. We were not expecting immediate results, but we were delighted to be proved wrong.

With brilliant simplicity, a single engineer chose to solve the pressing issue of helping customers choose less crowded trains, itself a consistent customer pain point.

The rail operators had previously attempted to address this via a complex multimillion-pound upgrade in floor-based pressure sensors. "But why not simply ask the customers?" the engineer proposed. After all, we were already operating at a significant scale, and with a small addition to the mobile application, travelers might just care enough to tell us.

And indeed, they did. Within weeks, the initial hack had been validated via a live-data prototype, generating hundreds of thousands of data points.

Our new data science team got to work, which allowed us to present the results back to customers. It was the first, critical example of our growing ability to harness data. And it also achieved a significant amount of free publicity—not just across national media, as nice as that was to see, but also in the halls of some very influential tech companies. Google's maps team reached out, and the Amazon AWS team took notice.

Engineering's significant efforts to lift and shift our platform quickly onto AWS had created a new strong partner. AWS was impressed by our expertise, and we were invited to showcase at that year's high-profile AWS conference, *re:Invent*.

Trainline headlined the event, and we were introduced by AWS CTO Werner Vogels himself. The solution born out of our hack day was showcased at the heart of our presentation.

We were providing real-time data across millions of rail journeys to help customers choose a less crowded carriage, all powered by unique user-generated data built on AWS. A number of high-profile engineers, who were among the thousands watching that day, opted to join the company. It was another example of how the impact of real transformation can propel a company forward in multiple unplanned ways.

Changing How We Decide What Problems to Solve

Trainline's mission, to make rail travel easier for everyone, had long been championed as an important factor in helping the UK achieve lower carbon transport choices. Indeed, many employees—me included—regarded that as a key reason for joining the company.

But a shared motivation, while helpful, is never enough to achieve meaningful alignment. With a consistent stream of talented new employees entering the building, it was clear that we needed to showcase a consolidated future. If we all pulled together and executed to our new, higher expectations, what might success look like? We needed a modern, strong product vision—one that would bind our shareholders, executives, stakeholders, and product teams together, as one.

To achieve that goal required a detailed understanding of our most pressing customer problems. Having already set some significant strategic business goals (international expansion, revenue diversification), the product vision would come from understanding and then solving numerous critical pain points for our customers.

A new head of product research joined with a significant skill set in deep customer analysis. Her team's goal was to produce a comprehensive and up-to-date understanding of our current and potential new customers, and how we could best achieve maximum value on their behalf.

Initially selecting small, representative cohorts in each of our various target geographies, we conducted multiple rounds of interviews, uncovering hundreds of issues. But we also saw major and immediate patterns—a small number of problems kept being mentioned over and over. We quickly gave these a name: the "Super Seven."

These seven problems appeared to be critical shared problems that cut across all customer territories and segments. Once we had validated our conclusions at scale, it quickly became clear: Solving even a handful of these issues might just be enough to propel us into a new era of growth.

But there was a problem. People thought of Trainline as a ticket company. Yet many of these issues were playing out downstream of the transaction flow, after the ticket had been purchased.

We were a rail-oriented e-commerce site, ingesting multiple rail APIs to help customers transact and buy tickets. Yet many of these newly uncovered problems existed beyond the transaction, such as missed connections, delays, or overcrowded trains.

Such issues had long existed in rail, but no one had ever achieved satisfactory solutions. Perhaps for good reason. They were complex problems, and we were unsure if we could help to solve any of them.

But we knew if we could solve these problems, then we could significantly improve the lives and commutes of our customers.

Centered strongly around mobile, our vision showcased an ambitious future. An early mobile application test, offering customers an ability to switch from paper to digital tickets, had been run. It resulted in a major insight: On these small number of mobile-approved routes, the frequency of rail travel and sales overall increased significantly.

The test had not been straightforward to run. In highly regulated environments, successful product innovation is frequently the result of significant amounts of complex work with partners, unions, regulatory bodies, and even governments.

We had a small but expert operations team that spent countless hours evangelizing with our rail partners. Little strong progress would have been achieved without them.

A shift to digital rail tickets would require our partners to make substantive changes to real-world working practices, underpinned by a multimillion-pound investment in infrastructure upgrades. Achieving this would require full governmental support, and a major update to the UK's transport policy.

But the data was clear. By allowing customers to skip the pain of ticket machines and more easily make changes and refunds, rail became a more relevant travel choice.

Like most dramatically successful tests, the result appears obvious with hindsight, but this was a substantial new data point: Mobile tickets increased the frequency of rail travel and, in doing so, increased customer lifetime value. This insight, alongside the associated deep understanding of how we could alleviate as much customer pain as possible, formed the core of a product strategy that allowed us to achieve a new and consistent path forward.

By now, we had perfected a solid quarterly cadence for the product teams, culminating in two high-profile days in front of the executive team at the end of each quarter.

Outcomes were front and center, and team members were significantly exposed to leadership: This was a format now enjoyed by most, even if it caused stress because outcomes are so much tougher to achieve than simply shipping output. But there was no going back. We would win or lose on the strength of the teams' ability to deliver results.

Our architecture had been reborn and was now a shining example of a cloud-native platform. The product teams had been empowered and reoriented toward strong outcomes.

Driven by highly collaborative product teams, we were testing hundreds of product ideas every year. Our data sets were now centralized, consistent, and useful. Our customer location data pointed at new possibilities to solve downstream delays and congestion. We were able to harness new and unique user-generated data, which was added to our already considerable journeys and pricing data. In parallel, we had begun to create our own machine-generated data—providing the ability to create ever more unique ways to solve a variety of entrenched, longstanding problems.

The Results

Trainline was now capable of doing things in the rail industry that no one had ever seen before.

One analyst memorably referred to us as the "Uber of Rail," a huge compliment at the time, given Uber's ongoing success and our late starting point. Given that Trainline was now the #1 travel app in the App Store from Monday through Friday, above Uber, we enjoyed the analogy. It represented a significant perception change at precisely the time when the market was deciding how to value our business.

A little more than four years after KKR purchased the company for below £500 million, the bell rang in London's Stock Exchange, and the company was valued at just over £2 billion, one of the largest European IPOs of the year.

The transformation had been successful. The power of capable product managers, designers, engineers, and data scientists—combined with our expert operational and legal teams—allowed us to dramatically increase our trajectory.

Problems to solve had replaced features to build; strong metrics had replaced supposition. Technology was now front and center.

Week by week, month by month, our visible success attracted ever more impressive talent to push us rapidly forward.

We had repositioned the company beyond rail, into the much broader (and more lucrative) travel category, also launching across multiple new geographies and even diversifying our travel options beyond rail into coach (bus).

Led by strong leadership, a deep understanding of the possibilities of strong technology talent, a highly collaborative culture, and, above all, a dedicated focus on outcomes, a small team improved the lives of millions of travelers and achieved an exceptional return for shareholders.

VI

The Product Model in Action

Thus far we've mainly discussed the theory behind the product operating model. In this part, we want to try to bring this way of working to life.

In the innovation stories, we shared what it's like inside a product team, but here we want to give you a glimpse into what this looks like as product teams interact with customers and with different parts of the company, including sales, marketing, finance, stakeholders, and executives.

There's one thing we want you to keep in mind as you read the chapters in this part: The chapters describe the interactions that you strive for. Since these interactions involve people, sometimes you'll fall short of your intentions.

The chapters here describe the case where the interactions are ideal. But even in the very best companies, with the very best people and intentions, things won't always go as you might like.

21

Partnering with Customers

In many companies that follow the prior models, especially companies that provide products to businesses, the relationship with customers is too often not where anyone wants it to be.

From the customers' perspective, the company may be perceived as unreliable and untrustworthy, incapable of delivering on its promises.

There are many reasons for this, but the bottom line is that the product model introduces a new, direct, and very different relationship between customers and the product teams.

As a company moves to the product model, there are some fairly significant changes to how the product teams need to interact with customers.[1]

Most of those changes are not directly visible to customers, as they involve how product teams do their daily work to discover and deliver solutions.

---

[1]We use the general term "customers" here to refer to the many different forms of customer that may be relevant—different types of users, buyers, approvers, influencers, internal colleagues using technology to serve end customers, and so on.

But as you know by now, the product model relies on direct and frequent interaction between the product team and the actual users and customers.

Partly this is to gain a deeper understanding of customers' problems and the environment and context in which a successful solution needs to operate, and partly this is to test your potential solutions to make sure that solution is valuable to customers, and usable by the various types of users.

We like to encourage product teams to explain their intentions as they begin to interact directly and intensely with users and customers, as this change may be very different from the customers' perspective as well.

Most customers are used to just dictating features they believe they need to their salesperson, and then they expect to see their features on a product roadmap at some point, hopefully soon, with a date they can start planning toward. It is also possible that the customers have already begun to lose confidence in the company's ability to deliver those features on the promised dates.

At some point, a frustrated customer may seek another company's product. But if customers currently are dependent on the company's products, for better or worse, they usually will try to hold on for as long as they can.

So, most customers are open to the idea of a change in the dynamics, but only if they believe that they're *more* likely to get their needs met rather than *less* likely.

The interactions that follow are intended to describe what customers can expect from an empowered product team working in the product model, and the reasons behind the differences. If you think these interactions are useful, you will need to decide if you would like to adopt these as they are, or modify them to suit your situation.

Promises

No longer do product teams promise a date or a deliverable until and unless they truly understand what will be required to deliver on that promise. This is very different than how you likely have been working, and this change has several significant implications:

The product team making that promise will need to engage *directly* with the relevant users and customers to truly understand what will be necessary to succeed. Note that companies often have well-meaning people assigned to manage relationships with customers, just as customers often have well-meaning people assigned to manage relationships with vendors. But these proxies—whether from your company or the customer—are not sufficient. Direct access between the product team and actual users is *essential*, and a prerequisite to any commitments.

The only people who can make a promise for a product deliverable are the product team that will be responsible for delivering on this promise.

Not the company's executives, not sales, not marketing, not customer success, not program or delivery managers, and not product managers. The promise must come from the *product teams*, including and especially the *engineers* on those teams, who need to fulfill that promise.

The product team will not make a promise until and unless they have conducted enough product discovery work to understand what is truly required, and whether that solution will be effective for the customer. This normally means one or more prototypes to address the specific risks. Once the needs are understood, the team will also factor in other work and commitments before they are ready to make any promise. We call these promises *high-integrity commitments*, and the product teams that make these commitments strive to do everything humanly possible to deliver on these commitments once made.

Strong product teams recognize that making commitments must be weighed alongside your other commitments, including potentially impacted keep-the-lights-on activities. What makes the commitment "high-integrity" is because it is based on first looking deeply at what is necessary, and then coming directly from the people who will need to deliver on the commitment.

The product team may need to remind the customer that it works for a commercial product company. Thus, the job of the product team is to come up with a solution that not only the specific customer believes is effective, but is one that will work for other customers as well.

This is the essential difference between a custom solution provider and a commercial product company. What this means in practice is that sometimes the specific solution that the customer has in mind is not general enough to work for other customers, and so the product team will need to explore alternative approaches. Most of the time, this is solvable and results in a better solution for all involved. But if it were to turn out that the necessary solution would work only in a single customer's environment, and the customer understands that it would not have the benefits of ongoing innovation and service, then normally the product team would instead direct the customer to a custom solution provider.

Product Discovery

The purpose of product discovery is to come up with an effective solution to your customers' problems. Your product team strives to discover solutions that your customers love, yet work for your business.

To do this, the product team has a cross-functional set of skills in product management, product design, and engineering. The product team engages directly with users and customers to deeply understand the problem that needs to be solved, and to test out potential solutions.

For the majority of direct customer interactions, the product team focuses both on understanding the customer's problem and on determining if you have come up with a valuable, usable, feasible, and viable solution.

As such, most of the direct interactions between the product team and the customer involve interviewing the different users and testing prototypes of solutions with those users.

Product Delivery

Once the product team has discovered and developed the necessary solutions, you have additional obligations to appropriately serve your customers:

First, you promise to effectively test your solutions, which means two different activities.

1. You will test that the new capability works as expected.

2. You will test that the new capability does not inadvertently introduce any problems elsewhere. These are called regressions, and while preventing regressions is difficult, and sometimes they occur despite best efforts, you also acknowledge you have a responsibility to ensure that new capabilities do not come along with unexpected problems.

You also understand that your customers urgently need the solutions you are building, but you also acknowledge that customers have their own jobs to do. Spending time recertifying your product and relearning and retraining on it is not something anyone wants to do, nor should they be required to do. You strive always to be sensitive to the cost of incorporating new capabilities into daily use, and you use modern design and deployment techniques to smooth adoption whenever possible.

Finally, you know you can't realistically promise that you will never make mistakes. However, you can promise to use best practices and best efforts to minimize those mistakes. And most important, you can promise that when mistakes are made, you will do everything in your power to correct the issue quickly and competently, and then analyze how you can avoid that problem in the future.

Happy, Referenceable Customers and Business Impact

Your product team's highest-order goal is to do everything you can to get each of your customers to the point where they are effectively using your products every day to do important and meaningful work, and for your customers to love and value your products so much that they happily share their experience with others. This customer success is then reflected in your business success.

This is how you feel about your favorite products, and you strive to work every day to ensure your customers also feel that way about your products.

CHAPTER
22

Partnering with Sales

Please note that this chapter does not apply to all companies and all types of products. But when it does apply—when the company has a direct or channel sales force responsible for bringing its products to market—the relationship between the product organization and the sales organization can literally make or break the success of the company.

There are perhaps no two more interdependent roles in a company than product and sales.

Product depends on sales to get its products into the hands of its customers. And sales depends on product providing them with solutions that truly meet their customers' needs.

If either one falters, we have a real problem.

Yet, despite this seemingly natural alignment in incentive, in practice they are too often pursuing very different goals. In so many organizations using the prior models, neither product nor sales is happy.

Product is not happy because sales keeps requesting (or demanding) they build things that they know are not going to solve the customers' real problems.

And sales is not happy because they are simply relaying the demands of the customers, yet product keeps delivering solutions that don't meet the needs of those customers or, worse, not delivering at all.

Realize that salespeople depend primarily on commissions for their income. This comes from new sales, renewals, upsells, and increased customer spend with your company.

For that reason, salespeople are wired to respond very directly to customer requests. Unless there's a genuine belief that the product team exists to support similar goals, the sales team believes they must battle for their very survival.

The good news is that the product model is designed to correct this issue and has a strong track record of doing just that.

First, it's essential to understand the necessary relationship between product teams and customers, as was described in chapter 21, "Partnering with Customers." The model of a salesperson collecting requirements from a prospective customer, and delivering those requirements to a product team to build, is not part of the product model.

However, at the core of the product model is the focus on creating happy, referenceable customers, and product needs to collaborate with sales to locate and develop these prospective referenceable customers.

It's worth pointing out that when product has not provided these happy, referenceable customers, the sales job is dramatically more difficult, in addition to fueling sales's distrust of product. It shouldn't be a surprise that this situation leads directly to requests for custom ("special") solutions in order for sales to be able to sell anything.

For the necessary changes to occur, the key is to build trust between product and sales, and the best way to do this is to spend quality time together in front of current and prospective customers. This starts with the head of product and the head of sales, but the goal is to spread this down to individual product managers and salespeople.

The salesperson is able to witness up close how product teams can discover a winning solution to customers' problems, often with solutions that the customers had no idea were possible.

The product team is able to witness up close how the salesperson navigates the many aspects of effective go-to-market to understand the buyer, the different types of users, the influencers, the approvers, and the other factors that play into both the purchase decision and successful use of the product.

Working closely together, the two major dimensions of product-market fit start to come into focus. Then working across multiple prospective customers, the product team learns both what is consistent across customers and what is not, so the product team can ensure it is able to create a single solution that will be successfully deployed across many customers.

None of the above is easy, but what makes it hard is not risk or complexity, but rather the effort needed by both product and sales. For those willing to put in the time, the rewards are real.

The product teams, with the help of product marketing, have a variety of techniques designed especially for these types of product efforts, and the goal is always happy, referenceable customers.

There is one more powerful consequence of the product model for sales: In most legacy models, the sales organization is the sole customer-centric group that is outcome based (performance compensated and incented on sales). The product model gives sales a true partner that is also customer centered and outcome based. When done well, product and sales are truly collaborative partners.

CHAPTER

23

Partnering with Product Marketing

One of the closest partners for product teams is your associated product marketing manager (PMM). It's not hard to see why establishing an effective partnership between product and product marketing is key to the product model working well.

In a few companies, the product team will be fortunate enough to be paired with a PMM (dedicated just to the product team), but in most cases, multiple product teams will share one or more PMMs. This is simply because it rarely makes sense to organize product marketing in the same way you structure your team topologies. Whether you have a dedicated PMM, or whether you share one or more PMMs, the product manager will want to reach out and establish the necessary relationship(s).

The nature and frequency of your interactions will range based on what you're working on at the moment, but there are quite a few different interactions where you'll need to collaborate.

While most people have at least an intuitive sense of the interactions with product marketing, many are surprised to see the depth and power of a strong and effective collaboration.

Here's what you strive for in the product model.

Market Understanding and Competitive Analysis

As product teams work to understand different markets, the PMM is a key resource with access to industry analysts, market insights, competitive research, and more. Moreover, as you learn about new competitors and new technologies, this information is shared between the product teams and product marketing.

A good example of a collaboration in the area of market understanding would be to evaluate a new competitive offering and put together a response that can be shared with the sales organization.

Product Go-to-Market

Product go-to-market is arguably the most important area of collaboration. When working on a new product for an existing market, or when taking an existing product to a new market, or especially in the case of creating a new product for a new market, coming up with an effective product is only half the problem. The other half is working out how to effectively get that product into the hands of customers, known as the product's *go-to-market*.

Your PMM partner is your resident expert on the many go-to-market alternatives and techniques, and when working to come up with product-market fit, product and product marketing need to work together to solve for the many constraints.

Once product go-to-market has been established, the PMM works to make sure each of the product teams understands the power and nuances of this path to market.

Even later, with incremental changes to the product, there may be messaging or positioning considerations based on the go-to-market where the product manager will want to consult with the PMM.

Key Product Decisions

Product decisions come up all the time in product discovery, and for many of those you will want to consult with your PMM partner. But for especially important decisions, because the PMM has such deep knowledge of the target markets and the product's go-to-market, the product team will want to be sure to include the PMM's perspective.

To be clear, a product's go-to-market is one of the most important considerations in determining successful product strategies.

Customer Discovery Program

The customer discovery program is a technique that is especially powerful for creating new products, or taking existing products to new markets, and this technique is designed to be a collaboration between the product team and product marketing.

You can learn more about this technique in the book *INSPIRED*, but it involves identifying, selecting, and working closely with a set of prospective customers from the same target market. Partnering with product marketing and sales, you interview and select these customers, and then proceed to discover and deliver a solution designed to meet their needs. The result is your initial set of referenceable customers.

The product manager and the product marketing manager work closely on all aspects of this program. The product manager applies the learnings to the new product while the product marketing manager applies the learnings to the go-to-market, sales tools, and sales process.

Messaging and Positioning

Even with day-to-day changes to the product, often visible changes have implications for messaging and positioning.

Sometimes these changes are tactical, and sometimes the messaging and positioning have broad implications for the product.

In this case, the product manager or the product designer may consult with the PMM to ensure alignment and that the new capability will be noticed, understood, and adopted.

Customer Impact Assessment

With modern product delivery techniques such as continuous deployment, changes are happening many times a day, but there are certain changes that may have an impact on your customers, your sales team, or your customer success team. In this case, you need to coordinate with your PMM partner to ensure that the upcoming change is not a surprise and that they are all prepared.

If the PMM partner is not prepared, maybe because they only just found out about the change, then either you would hold off on the changes, or if you have the suitable deployment infrastructure, you could release the new capability in such a way that it is not yet visible to customers until the PMM reports that they are ready.

Pricing and Packaging

In most companies today following the product model, the product teams defer to product marketing for pricing and packaging decisions.

This is due partly to expertise (many product marketing groups have relationships with specialty pricing firms), and partly because pricing is usually done at a higher granularity than the product team, which is more aligned to product marketing.

Product teams certainly provide input to the pricing decisions, but many other factors are used to calculate the ideal price.

Sales Enablement

The various product teams each contribute expertise to the PMM, but the PMM is responsible for the product's collateral and the various sales tools used to equip the sales force with what they need.

If product managers have spent time developing the relationship with the PMM, hopefully the product teams feel good about the PMM's level of understanding of the product's capabilities and nuances. Understand that if this is not the case, then much of this work will fall on the product managers. This is why experienced product managers learn that spending time explaining nuance to the product marketing manager is often a very high-leverage use of their time.

For more information on the critical role of product marketing, see the SVPG book: *LOVED, How to Rethink Marketing for Tech Products* by Martina Lauchengco.

CHAPTER

24

Partnering with Finance

Partnering with finance, and often with the related investor relations, is a critical part of the job because there is simply no viable alternative to working together effectively. The company depends on the products it sells, and product people depend on finance.

As with so many other areas, the key to working together effectively lies in alignment, understanding the needs and constraints of the partner, and providing data rather than opinions.

First, it's important for the head of product to establish with the head of finance the reasons for the move to the product model.

Start by looking together at the data and acknowledging that the current model is not very predictive in making good funding decisions, and not very predictive in estimating incremental revenue from product. If there's doubt, you can see this by comparing the business case claims with the actual results from the prior quarters.

More generally, it's important to acknowledge that the old model simply wastes too much money and time, not to mention opportunity cost, and generates far too little in the way of demonstrable business results.

Usually this much is painfully clear. The only real question is whether we can do better.

Testing the Product Model

The case to be made is to test out the product model, and learn together if it can provide the improved returns that it has provided to other companies.

Depending on the level of frustration with the old model, and the level of urgency to get to the new model, you can test out the new model either very conservatively (with a small number of product teams) or more aggressively.

It may also be useful to point out that if the move to the product model succeeds, then in addition to happier customers and better business results, the company might also get valued by public markets at least in part as a tech-powered company, which is often a motivation for investors.

Product and Finance Collaboration

As part of moving to the product model, a number of interactions would change in how product and finance collaborate:

The product teams promise to stop hiding behind features and roadmaps, always promising better results once the next new feature is ready. The product teams will stop making up business cases and then making up excuses when the results don't materialize.

Instead, product signs up to be measured just as your company is measured, by business results. If the company launches a new product, it will be measured by business success. If the company adds a new capability, it will be measured by value and business impact.

If a product team signs up to solve a customer or company problem, it will define success by business results, and the team will work until that result is achieved, whether that means one new feature or many features or a different approach to the solution.

The product team will admit what they don't know, and tell you honestly what they can't know. When an answer with integrity is needed, they will run the necessary testing to get the data to make an informed business decision.

Going forward, when they bring you a request for incremental resources, they will bring you data, and a transparent analysis with the reasoning for that request.

Rather than large funding requests, the product team will first run very low-cost tests to collect the necessary data. Then, if they have a larger request, it will be accompanied by the actual data that supports this investment.

In the cases where the company needs a date that can be counted on for a deliverable, the product team will use a process to deliver a high-integrity commitment. While this takes somewhat more time, the date will be defensible, and something against which teams can be held accountable.

When the product model starts producing real and consistent innovation, product will remember to share with finance and investor relations the evidence that these investments are paying off with real results.

Requests of Finance

For this partnership to work, there are a few things that product needs from finance:

- Rather than make funding decisions on a project basis, product asks that you staff product teams with specific focus areas and measures of business results, for one or more quarters. The product teams will provide real-time business results with results judged quarterly.

- The product model requires several new job competencies. In most cases, existing employees can be coached and trained, but in the event this is not possible, product leadership will provide a proposal for replacing existing staff with new staff. It is worth noting that in most cases, moving to the product model saves money rather than costs money, even though this change may involve insourcing critical competencies.

- Rather than being judged on shipping features and projects, the product teams ask to be judged on business impact. To do this, the product teams need the room to run the necessary tests and make decisions about which features and projects will best meet the objectives. The intention is to leave this level of granularity to the product teams.

- Finally, the product organization asks finance to minimize the need for high-integrity commitments, as they represent nontrivial work just to provide the date, and they can be disruptive to the organization.

When product and finance work together effectively, they can get the most from their investment dollars, as well as manage the financial risks more responsibly and effectively than ever.

CHAPTER

25

Partnering
with Stakeholders

We define a stakeholder here as someone not explicitly a member of a product team, yet who represents a key constituency, area of the business, or special expertise.

There is no denying that moving to the product model represents a big change for the business's stakeholders.

Some of the stakeholders are so frustrated with how things have worked in the past that they are desperate to try something new.

Some will feel they still have the responsibility to deliver for the business, but now without the control over the technology resources they once had.

Others will take a wait-and-see approach.

But in all cases, since in the old model the technology resources were there to serve their needs, and now those resources are there to serve end customers directly, they realize there's a significant change involved.

The product model is designed to have the product teams as collaborative partners with the stakeholders. Yes, they are no longer subservient to the stakeholders, but they do still depend on them.

In practice, this means building trust between the stakeholders and the product managers.

An empowered product team is designed to solve problems for your customers or your business, in ways that your customers love, *yet work for your business*.

Creating a solution that customers love is usually not that hard, assuming the product team has direct, unencumbered access to the users and customers, and the product data.

But creating a solution that the customers love *and* that also meets the many, often-competing needs of the different parts of the business can be very challenging.

You need to ensure your solutions can be marketed, sold, and serviced effectively. You need to ensure you can fund the product, and you can then effectively monetize that product. You need to be able to operationalize the product. You need to ensure the product is compliant with relevant regulations, respects privacy laws and partnership agreements, and does not have unintended consequences on people or the environment.

We know you can't do all this alone.

But we also realize that to create any product, literally hundreds if not thousands of decisions need to be made. Bringing all the relevant stakeholders together for every decision is not only an excessive use of time and money; we also know committees very rarely innovate.

To innovate, you need direct access to customers, and direct access to the enabling technology.

We also recognize that for a stakeholder to build a respectful, trusting relationship with a product team, *the onus is on the product leadership to ensure that every product team has a competent product manager who has put in the work to understand the various constraints of the business and to be an effective partner to the stakeholders.*

Expecting stakeholders to trust a product manager who does not have this foundation is not only unrealistic, but also unwise.

Yet even though product managers are on the product team to represent these various business constraints, very few product managers know all the relevant business dimensions in sufficient depth.

Therefore, a product team's commitment to stakeholders is that each product manager promises to engage directly with each of their relevant stakeholders and work to learn and understand the various constraints and needs represented by each area.

Further, product teams realize that it takes time to build this knowledge and this trust. Whenever they are considering a solution that *might* impact a particular stakeholder, the product manager commits to show a prototype of the proposed solution to the stakeholder so that they can consider the implications of this change *before the team builds anything*.

Hopefully, if the product manager has correctly understood the various constraints, the stakeholder needs only a few minutes to take a look and give a thumbs-up.

However, in the event there is a problem, the product team can quickly iterate on the prototype until the stakeholder believes the solution will now address their concerns and work for the business.

Sometimes, this will be easy. Other times, getting a solution that works not only for customers, but also for multiple stakeholders—each with different needs—may require many iterations to identify a solution that solves for all parties. Further, the solution that solves for all parties may end up being something that none of the stakeholders had imagined was possible.

More generally, the product manager strives to work toward the desired business *outcomes*. The product teams know that shipping features won't necessarily solve the underlying problem for the customer or the business. The product team is committed to pursuing the necessary *results*.

But the promise is not to build—and certainly not to ship—any solution that does not meet the critical needs of the business.

In the event that something that is not viable for the business does manage to get built—or, worse, gets shipped to customers—the product team promises to correct the situation as quickly as possible and to determine how the error occurred, so it is not repeated.

The product managers promise to be respectful of stakeholders' time, and to earn their trust. They also strive to be transparent. Stakeholders are welcome and encouraged to participate in testing with users and customers. They are welcome to look at any of the prototypes created. They are welcome to view the data on the product—both how the products are used, and the results of live-data tests.

Working together, product teams and stakeholders can solve problems for your customers and your company in ways your customers could not have imagined possible.

CHAPTER

26

Partnering with Executives

Depending on the current culture, moving to the product model may represent a significant change in how the executives interact with product teams and product leaders.

Moving to the product model often requires significant cultural change, especially when moving from top-down, command-and-control leadership styles.

When we discuss the necessary changes, we mostly focus on the changes needed within the product and technology organizations. However, an important aspect of transformation is changing the dynamics and interactions between the senior executives and the product organization.

It's true that many product leaders and product teams would like to think this is as easy as saying to the executives "Please just back off and give the product teams space to do their work."

But this ignores the reality that executives have very real needs when it comes to running the company responsibly and effectively.

And to push decisions down to product teams, those product teams need to understand the strategic context, much of which comes from the executives.

So, rather than reducing the level of interaction, in the product model, product teams need frequent, high-quality engagement.

What we're really talking about is the *nature* of these interactions.

You hope to interact in ways that provide executives with the information they need to run the business, yet empower the product leaders and product teams to do the work they are capable of.

We've found that a set of techniques can help encourage constructive and effective interactions between the executives and the product leaders and product teams.

Decisions

One of the central ideas behind the product model is to push decisions down to the product leader or product team that is in the best position to come up with the best answer. Normally, those are the people working directly with the enabling technology, and directly with the users and customers.

But teams can make good decisions only when they are provided the necessary strategic context. So, it's critical for executives to share the broader context: the business strategy, financial parameters, regulatory developments, industry trends, and strategic partnerships.

Product teams depend on the executives to share as much of the relevant strategic context as possible, so they have the information to make good decisions.

And executives depend on product leaders and product teams to share openly and honestly the data and reasoning used to make their decisions.

Outcomes

Product teams understand that what matters is results. Whenever possible, product teams expect to be accountable for *outcomes* rather than *output*.

But this works only if product teams are given *a problem to solve* (rather than a particular potential solution to build) and are then empowered to come up with a solution that works.

To come up with the best solution they can, product teams depend on executives to provide as many degrees of freedom as possible to solve the problems they're assigned.

Product teams then promise to take responsibility for coming up with a solution to the problem, and make best efforts to simultaneously solve for the customers, solve for the business, and solve for the technology.

Disagreements

Product teams realize that the executives have the broadest access to data and the best understanding of the wider business context. Yet product teams have access to the enabling technologies and to the actual users and customers interacting with the products every day.

Product teams realize that at times there will be differences of opinion. In the old models, contrarian or counterintuitive ideas are not explored and innovation rarely happens. In the product model, very often these counterintuitive insights power real innovation. If something is important, you coach product teams on how to run tests quickly and responsibly to collect the necessary evidence, or where necessary, proof.

Product teams depend on executives to allow for and even encourage teams to explore alternative approaches when they believe that's necessary to succeed.

If a product team proposes an approach with associated risk, the team promises to responsibly run product discovery tests to collect the necessary evidence and then share those findings.

Promises

Product teams understand that there will be times when an executive needs to know a specific date for a specific deliverable. The teams commit to treat these promises as *high-integrity commitments*.

Product teams understand the damage that is done to an organization when promises are made and then not kept.

Product teams ask that only the product team responsible for delivering on a promise be the one to make that promise, and teams will not be asked to make a promise or deliver on a commitment where they don't know what is involved and what would be required to succeed.

Further, product teams depend on high-integrity commitments being the exception and not the rule, as the time and effort in making and then delivering on a high-integrity commitment can be significant.

But once a promise is made by a product team, they pledge to then take this high-integrity commitment very seriously and do everything within their power to deliver on it.

Surprises

Product teams realize that surprises are sometimes unavoidable, but they also know they must try hard to minimize them. This means that if there's anything even potentially sensitive, the product manager is expected to recognize this and to show and discuss it with the impacted stakeholder or executive *before* the solution is built.

One of the worst forms of waste is when a product team builds a quality, scalable implementation of some capability, and finds out afterward that there is some reason why this is not an acceptable solution.

Product teams commit to preview any potentially risky solution with the relevant executives before building it.

Similarly, if an executive reviews a prototype during discovery and doesn't raise any major issues, but once the product is built and in production decides that something is now a serious problem, then this is likewise very wasteful and frustrating to the organization.

Regardless of the cause, when something gets built but is then deemed to have a serious problem, it is reason to have a postmortem discussion to see how that type of waste can be avoided going forward.

Trust

Product teams understand that empowerment depends on a degree of trust, and they strive to earn that trust.

Likewise, product leaders and product teams trust that the executives are leading the company in a positive direction.

Neither expects the other to be perfect, but both product leaders and product teams each recognize that they depend on the other and commit to best efforts to help each other succeed.

27

Innovation Story: Gympass

Marty's Note: There are terrific examples of transformation and innovation from all over the globe today, and I love what this Brazilian scale-up was able to accomplish when they were put to the test by the pandemic.

Company Background

Gympass is a Brazilian company started in 2012 with the mission to improve employee health and well-being by helping to defeat inactivity.

Gympass enabled companies to provide their employees with access to a network of more than 50,000 gyms and studios in 11 countries through the Gympass platform.

For the first five years of the company's existence, as it grew from an early-stage startup to more than 800 people, there was only a very small IT organization. The company mainly depended on a large business operations team largely running the business using spreadsheets.

To leverage technology effectively, in 2018 the company decided it needed to transform, and the leaders brought in experienced product leader Joca Torres as their chief product officer.

Working with their CTO and CMO, the three leaders rapidly built out a technology and product organization. The teams began their work by automating most of the tasks done by the business operations team, as well as creating all new experiences for the gyms, the end users (the employees), and the HR staff (so they could see the impact on their employees).

These efforts paid off quickly, and the company was on a strong growth trajectory going into 2020.

Unfortunately, then the pandemic hit.

One of the industries that was severely and immediately impacted by the pandemic was the fitness industry, especially in-person gyms and studios.

Not only were employees often not able to return to the office, they couldn't go to their gyms, either.

Countless gyms and fitness industry companies were forced to shutter their doors.

As with so many companies during this time, the leadership team at Gympass was faced with an existential crisis. They knew that their mission of defeating inactivity was now more important than ever, but they also realized they would need to pivot their approach meaningfully and quickly.

The Problem to Solve

Fortunately, the company's product organization had been learning and testing many fitness-related product ideas with their users, and they had already discovered that a significant portion of users were interested in wellness activities beyond gym-based fitness, such as meditation, mindfulness, nutrition, and home-based exercise.

The problem the product teams needed to solve was to provide a range of home-based wellness solutions to Gympass users so they could maintain their fitness, even without access to a gym or studio.

Strategically, the product team took a build-and-partner approach—building key parts to the solution but partnering with other fitness-solution providers for specific vertical fitness solutions.

Discovering the Solution

The team began an intense product discovery effort consisting of a rapid series of prototypes intended to address what they considered to be the two major risks:

The first and most serious risk was to ensure that the solution would be *valuable* to the end users. If their end users (the employees of their client customers) would not choose to use the new solution, Gympass knew it was only a matter of time before the companies would stop paying for this benefit.

The second was to ensure that this new product and associated business model would be *viable*—especially that there was sufficient revenue for the fitness partners and, of course, for Gympass—yet still complied with the client company's benefits requirements.

The main tool for this discovery work was a user prototype that demonstrated the user experience with the various wellness services. There were also several feasibility prototypes testing out the necessary partner integrations.

Once the team felt they had addressed these risks, they needed to move from prototype to product in record time.

In addition to the product development and delivery, the company recruited nearly 100 vertical fitness partners in less than two months, so they could be sure to cover the range of wellness activities in the many languages of their users.

The Results

Gympass rolled out to ten countries within four weeks and went from zero to hundreds of thousands of users in just a few months.

The product team that did this work consisted of one product manager, one product designer, and four engineers collaborating with a dedicated business operations person and a product marketing manager. The sales organization also helped by recruiting local partners.

If Gympass had not invested in the effort to transform, they almost certainly would not have had the necessary skills to so rapidly pivot the business, and they would very likely have suffered the same fate as so many others in the fitness industry.

Instead, during the course of the pandemic, Gympass managed to more than double the number of corporate clients and their revenue. Company valuation rose to US $2.2 billion, resulting in one of the success stories of the Latin American tech industry.

VII

Transformation Story: Datasite

By Christian Idiodi

Marty's Note: When I first learned Christian was going to join this company as their head of product, I had already known him for several years and considered him one of the industry's top product leadership talents. I told him I could get him in as head of product at almost any marquee company, but he really wanted to see if he could transform an old, sales-driven, financial services company to become like the best. Even though I knew he was good, I told him that he was picking an exceptionally tough challenge. But if you know Christian, you know that he is not intimidated by difficult challenges. And he joined with two other proven leaders: Thomas Fredell and Jeremiah Ivan. The three of them managed to redefine what is possible.

Located in Minneapolis's industrial district, Merrill Corporation was founded in 1968 by Roger Merrill and his wife. When I arrived,

Merrill was an organization in the midst of an identity crisis, and it desperately needed to change.

Merrill was a traditional, family-owned, brick-and-mortar operation primarily known for its financial printing services, which included the printing, formatting, and distribution of financial documents, such as annual reports, prospectuses, proxy statements, and other SEC filings. Merrill also provided document management services to help clients organize, store, and share information. Included were solutions for document capture, indexing, storage, retrieval, and distribution. Other services included virtual data rooms that allowed clients to securely share confidential documents and information with authorized users.

But, of course, the world had changed.

The company knew it needed to transform and, in fact, had already tried outsourcing transformation and failed miserably.

Moving forward, the CEO Rusty Wiley knew Merrill needed to do this itself, so Rusty brought in Thomas Fredell, Jeremiah Ivan, and me, and the three of us got to work.

The Motivation

An array of external factors inhibited Merrill's ability to successfully compete in this rapidly changing marketplace:

- *Heavy reliance on legacy business.* Merrill Corporation was primarily focused on providing printing and communication services to the financial industry. However, with the increasing adoption of digital technologies and the decline in traditional print media, Merrill's business model was quickly becoming outdated.

- *Limited technological innovation.* Merrill lagged behind its competitors in terms of updating its existing technology as well as bringing new products to market. Part of this deficiency stemmed from the company's inability to understand its customers' evolving needs.

- *Legacy model.* Merrill had a project-based model for executing tasks, combined with a command-and-control approach that limited internal motivation and innovation.

- *High debt levels.* The company had a significant amount of debt, which put pressure on its financial performance and limited its ability to invest in new growth opportunities.

- *Reputation damage.* Merrill Corporation faced several lawsuits and regulatory investigations related to data breaches and information security issues. These incidents damaged the company's reputation and undermined customer trust.

These were the most pressing problems Merrill was experiencing when we joined the company. But I knew they were just symptoms, and I needed to identify the root causes.

A Slow-Moving Train

For the people at Merrill Corporation, the reason they had fallen behind was simple: Competitors appeared who were faster, less expensive, and knew how to leverage technology.

And they were right, but only partially. To understand the underlying problems, I spent a lot of time with the head of sales, Doug Cullen, who turned out to be a trusted ally. He told me: "Christian, what you need to understand is that Merrill is, first and foremost, a sales-driven company, which means that our culture derives from that. Because the salespeople own the clients, they also own everything else. If we decide to solve a customer problem, it's only because the salespeople want it solved. If we decide to build a new piece of technology, it's only because the salespeople want it built. Sales takes the lead in everything, and everyone else simply follows."

To underscore this point, Doug told me about the seventh version of a new technology platform the company had been developing and the software bought to try to compete with new competitors.

The platform project was managed by a director of sales who had outsourced the development to a firm in India. The platform had taken close to six years to complete at a cost of over $6 million. When it was finally released, it generated less than $30,000 in revenue.

So, one root cause was clear: The power center was within the sales organization, and technology had a subservient role. This creates all sorts of competitive weaknesses.

For example, salespeople—driven by personal quotas—will often prioritize their specific opportunities over the company's strategic goals.

Different customers can receive different views of the company and different levels of service depending on which salesperson they interact with, which can negatively impact the company's reputation as well as customer loyalty.

Another problem is that salespeople who are focused on their specific deals and opportunities are not well positioned to identify new market or product opportunities across the customer base.

This leads to a lack of innovation and makes the company vulnerable to competitors who are more effective in identifying and addressing customer needs.

A Disempowered, Mercenary Culture

The primacy of Merrill's sales-driven culture enabled me to quickly identify another of the company's core problems: disempowerment.

With an abundance of heroism in the sales and service organizations, other parts of the company were treated like mercenaries. Engineering was outsourced, decisions were top-down, and teams were given roadmaps of features and projects to deliver.

Merrill's office space, particularly the high cubicles and lack of sunlight, generated a feeling of isolation and disconnect among its team members.

This made it difficult, if not impossible, for its employees to collaborate and communicate effectively, further reducing their sense of empowerment.

Add to this a hierarchical floor layout, as well as a lack of shared spaces, and the result was low morale—and even lower engagement.

Because Merrill's employees did not feel empowered to make decisions or take ownership of their work, they were unwilling to take risks or suggest new ideas, for fear of rejection.

Because innovation and creativity were stifled within the organization, the company's ability to adapt to changing market conditions was essentially nonexistent.

A Lack of Focus

With more than 4,000 employees spread across 23 countries, Merrill Corporation had grown significantly over the years—not organically, but by acquisition.

While Merrill's portfolio of assets was profitable overall, there were no unifying synergies or theme. The sales organization didn't have a bigger picture to sell to customers. Instead, the company's sales were heavily dependent on personal relationships and service.

Merrill simply had too many balls in the air. Without a defined vision and strategy, the company lacked a clear direction for where it wanted to go and how it planned to get there. It was difficult for employees to understand what they should be working on and how their efforts contributed to the company's overall success. It also was nearly impossible for Merrill to decide which products or services to invest in, which markets to target, and which initiatives to prioritize.

Without any focus across its core businesses, Merrill had begun to lose market share. A real wake-up call arrived in the form of a new competitor called Workiva, which offered customers a modern, cloud-based solution.

As one customer after another began to use Workiva to file their quarterly reports and statements, Merrill realized that if it wanted to remain a viable force in the marketplace, it had to change. But what did change mean?

Transforming to the Product Model

We knew we needed to change how we worked at every level, from how we build, to how we discover, to how we focus.

For a company as old as Merrill Corporation with its ingrained sales-driven culture, this meant a deep and profound change.

For companies that were "born digital," technology is not separate from the business; technology *is* the business. For Merrill Corporation, the exact opposite was true.

Merrill was a sales-driven company. Revenue was generated and managed by the sales team, so the role of technology was to serve the needs of the sales organization.

For the new model to be successful, technology could no longer be subservient to sales. Instead, it had to move to the forefront where its primary role was to serve the customer—as both a solver of problems and an enabler of solutions.

Changing How We Build

As a result of age, long-term underinvestment, and the heavy use of external contractors, the legacy platform was fragile, obsolete, and simply unable to support the demands of the market.

The company understood that it had to reinvest in a more modern platform that enabled a faster response—not only to problems, but also market opportunities and competitive pressures.

For Merrill, the solution was to move from the existing legacy systems to a modern, microservices-based architecture.

Many legacy companies have this problem, but it's also true that many of these companies do not have the skills to dig out of this hole. Fortunately, when Thomas recruited me to run the product organization, he also recruited an exceptionally strong head of engineering, Jeremiah Ivan, whom we had both worked with before.

We needed a platform that not only could support the needs of our customers today, but could enable the necessary rapid experimentation and deployment of new capabilities that we knew would be required going forward.

The Need for Speed

It's also important to note that, for Merrill, the goal of changing to a modern platform wasn't simply about achieving faster deployment, but also the ability to find and fix problems quickly.

To underscore how important this was, I'll relate a story that could have had a catastrophic ending but turned out to be an inflection point for the transformation.

Early one morning, we received a call from one of our biggest clients. The client had found a security bug in our system and was irate. The mistake was so significant to the client that it was threatening to cancel its contract with our company. Alarms went off, and the call escalated to my desk.

I immediately reached out to the head of sales and the CEO and told them to get ready for a call from the client firm's CEO. At the same time, I put in a call to our VP of engineering and told him about the problem.

In less than an hour, the engineering team had located the problem and implemented, tested, and then deployed a fix.

At 10:00 a.m., we received a second call from the client—not to complain but to apologize.

"We thought there was a problem with the system," he explained, "but apparently we were mistaken. So, you can forget about the call to your CEO. Everything is good."

When I relayed the client's second call to Rusty, he was relieved, but also excited. He'd been making investments in technology and people but, from his perspective, had yet to see real benefits. That now changed.

"Christian," he told me, "the new muscles within the company to identify and diagnose a problem are impressive. Our ability to resolve that problem is also impressive. But what really stands out for me is the fact that our engineering team was able to respond to our customer's needs as fast as they did."

Rusty was right. The importance of this fix was significant not only because it solved a real customer problem, but because it was emblematic of all the technology and process changes the company had been working so hard to implement. These included:

- Changing from on-premises to a cloud-based platform
- Changing from a waterfall to an Agile delivery model
- Changing from a monolithic platform to independent microservices
- Changing from infrequent occasional releases to frequent, small releases (CI/CD)

This new optimized environment increased availability while improving stability, with a guaranteed service-level agreement. The solution also reduced the company's operational costs by 30 percent. The new platform provided access to dashboards, and clear visibility of operations, and enabled detailed infrastructure and database monitoring for all the company's critical applications. If more capacity, new storage, or additional computing power was needed, it could be quickly provisioned.

Changing How We Solve Problems

Very often success is determined not by our plans but by the people and processes we have put in place that enable us to respond to uncertainty and surprise.

Rather than simply approving and funding a six-month plan and then blindly proceeding to execute on that plan despite everything that is learned during that time, we instead fund an empowered product team to solve a problem and equip them with the skills to be able to adapt and learn from the customers and the technology every day.

Collaborative Problem Solving

Problem solving is a collaborative process. It's not the customer on the front lines, or the salesperson, or the all-knowing CEO.

None of these people know what is just now possible. For this reason, most innovative solutions come from those working closest to the technology.

But for that to happen, teams need to be informed and empowered to make decisions. They need a set of competencies and skills to tackle the risks involved when identifying and delivering a solution worth building.

Top-down decision making is not only slow and cumbersome, but the decision is too often made far from where the critical knowledge of the customers and the technology resides.

Responding to Customer Needs

Empowered product teams aren't concerned only with technology questions. They ask customer questions about users, customers and markets, needs and motivations. They ask business questions about go to market, compliance, security, privacy, costs, and monetization.

More generally, if they are creating new solutions, these product teams want to know if people will actually buy and use them before a single line of code is ever written. They also want to know all the risk factors to ensure that the products they create don't break—that they don't hurt our customers, our colleagues in sales and service, our company's reputation, or our revenue.

How did we empower the product teams?

The first step was to move from funding projects to funding people and teams. Doing this entailed giving them problems to solve rather than features to build.

We then created a safe environment in which team members knew they had the authority to make decisions without finger-pointing or micromanaging from above. We also staffed those teams with problem solvers—people willing to sign up to deliver business results.

This was a significant part of Merrill's transformation. The company had to shift from its legacy project-based funding model and direct its budget toward problem-solving teams.

It meant staffing new roles, such as product managers and product designers. It also meant moving from outsourcing engineering to building an engineering competency within the company.

What these collaborative teams did was enable the company to discover solutions to real customer problems in ways that our customers loved, yet worked for our business.

Leveraging Joint Knowledge and Experience

Deciding to transform to the product model doesn't automatically make you a better problem solver.

That's what empowered teams bring to the game: Smart people with different skills working together to solve problems that result in better products.

Finally, creating empowered teams doesn't mean taking power from one group and handing it over to another. Instead, it means bringing together people from different areas and leveraging their joint knowledge and experience.

At Merrill, one of the ways we did this was by allowing our product and technology people to talk with customers.

And changing how we solved problems meant:

- Eliminating the legacy roles of business analysts and product owners
- Hiring and coaching professional product managers
- Staffing a true product design competency—hiring professional product designers
- Moving from outsourcing engineering to building an internal engineering group
- Creating a "troika" model of product, design, and engineering, where a product manager, product designer, and engineering tech lead work collaboratively, with sales and marketing, on a solution to a problem they have been assigned
- Building a product discovery competency

Introducing product discovery and the critical product competencies represented adding major new skills and muscles to the organization and required a commitment from the leadership team to work together to raise the skills of the organization.

Changing How We Decide Which Problems to Solve

When companies say that *everything* is important, that they have too many priorities, the standard excuse for not getting anything done is to point the finger. They say, "We have a resources problem" or "We have a technology problem," when the truth is what they really have is a *focus* problem.

Focused problem solving is the result of a compelling product vision and an insight-driven product strategy.

The product vision should be big enough to last us several years, but we can change the product strategy as frequently as we like— usually quarterly—based on what we learn.

Context and Insight

The best decisions for which opportunities to pursue or issues to address are informed by context and insight.

Insights tell you *what's* happening. Context tells you *why* it's happening. Insights come from customers and competitors, from markets and industries, but, most important, from our data.

Context comes from knowledge and experience. It applies perspective to the data. Combining insights and context gives you the full-screen picture of what is happening, why it's happening, and what needs to be done.

This is what the product model is really about. The decision as to what opportunities we pursue is driven by the context and insights, not by reactionary or opportunistic business requests.

For Merrill Corporation, this meant working from a unified product vision and having a well-articulated product strategy.

We had to be clear on which customers we wanted to serve and what problems we wanted to solve for those customers.

And this meant all the things we were not going to do. The result was divesting (selling off) several businesses and shutting down others, so that we could truly focus on the best opportunities.

The Results

In 2020, Merrill Corporation rebranded itself as Datasite. The new identity was meant to reflect the completion of the company's transformation in which it divested its legacy financial printing business to focus on its rapidly growing, global, SaaS-based technology platform for the mergers and acquisitions community.

Before the transformation, the old Merrill Corporation could take months if not years to decide on, build, and deliver new capabilities to its customers.

The new Datasite is pushing the envelope to the point where it is introducing new value to its customers virtually every day.

All this speed and innovation has translated to the bottom line. In 2019, company revenue increased by more than 30 percent as it facilitated more than 10,000 M&A deals annually. Today, Datasite is not only the industry leader, but is highly admired for its culture of innovation.

VIII

Transformation Techniques

Now that we've discussed what it means to transform to the product operating model, the new product model competencies, and the new product model concepts, and we've shared several examples of companies that have successfully transformed, we are ready to discuss the more general topic of change management.

In other words, what is the best way of getting from where you are now to where you want to go?

Depending on the size of the organization working to transform, it can take anywhere from six months to two years to move to the product operating model, and that's assuming the company is serious about the transformation and it's not just theater.

The seven chapters in this part are all about sharing the techniques we've found helpful for working through your transformation.

- Transformation Outcome
  We'll start with our objective. When will the transformation be done?

- Transformation Assessment

  It's important to start with a clear understanding of just where the company is today.

  We'll share with you an assessment tool we've developed that we use when we engage with companies to help them transform.

  The assessment starts with a high-level, holistic view of how products are produced and then dives deeper into the specific product model competencies, concepts, and principles.

  For each item, we'll share with you what to look for.

- Transformation Tactics

  Many different tactics can help you in your transformation journey. We have organized them into tactics that help establish the product model competencies, tactics that help establish the product model concepts, and tactics that help your organization learn the product model.

- Transformation Evangelism

  Next, we'll discuss the importance of consistently and frequently evangelizing the product model and the progress with the broader organization.

- Transformation Help

  Finally, how do you teach the product model to the members of your product teams? We'll discuss what to do in the case where your leaders are already experienced in the product model, and in the case where they have not yet worked in the product model.

28

Transformation Outcome

When will your transformation work be done?

At one level, strong product companies are continuously improving, so in that sense transformation is never really complete. However, there is one very meaningful milestone that matters.

In the product model, one of the product model principles is that product teams are accountable for *results*. It's not enough to roll out new techniques. It's not enough to train your teams. It's not enough to produce output. Product teams must produce *results*.

If a product team doesn't produce real results for your customers and your company, then what have you really accomplished?

This is why it's so important when discussing transformation to the product model that you always frame this work as being about delivering real results.

This is also why throughout this book we share so many stories of real innovation by companies that transformed to the product model.

In every case, these companies simply didn't have the ability to generate those results without first developing these new muscles. This is the real reason for moving to the product model.

So, when we talk about the outcome of transformation, we like to frame the discussion around results. What do you hope to be able to do once you transform that you can't do effectively today?

Because you can't predict the future, you can't say what your specific goals are for the opportunities that present themselves in the future.

Ultimately, most companies want to be able to identify and take advantage of the most promising opportunities, and to respond effectively to the most serious threats.

Many companies that had earlier transformed to the product model were able to adapt to the challenges of the pandemic and come out stronger than ever.

Companies that had earlier transformed to the product model were able to quickly learn relevant new technologies, including the new generative AI technologies, and solve problems for their customers in ways they never could before. They are now stronger than ever.

No one can predict when a major new enabling technology will become available. All you can do is prepare yourself and your organization to be as ready as possible to deal with the threats and exploit the opportunities.

CHAPTER

29

Transformation Assessment

Overview

An honest and accurate assessment of the organization's current situation is essential to any plan to successfully transform.

To set your expectations, an experienced product coach usually can assess an organization (up to a single business unit) in as little as one day. The advantage of experience is to know whom to ask, what to ask, and what to look for.

But the intent of this chapter is to help you do this assessment yourself, even if you've never done anything like this before. In this case, the assessment likely will take a few days.

Before we jump right into the assessment itself, there are some very important caveats:

Be Realistic

No company is perfect. In even the very best product companies, you can occasionally find product teams that are not working like their

189

colleagues, or that have a few managers who revert to top-down, command-and-control leadership styles when under pressure.

Usually, they are very aware of this situation and are working to correct it, but that doesn't mean that the company as a whole is not working in the product model.

Too many product people are naive about this and have unrealistic expectations, and these people quickly lose credibility with the senior leaders. So, when you assess a competency or a capability, you are not looking for perfection—you are looking to understand the majority case.

At the other extreme, it's rare to find a company that is truly terrible at everything. There are usually at least some pockets of good, and it's important to recognize and acknowledge those.

Talk to All Levels

To get an accurate picture, it's essential to speak with people at all levels of the organization, from the CEO/GM down to individual engineers. It is remarkable how perceptions and understanding can vary, and also how some managers—especially middle managers—are so effective at obscuring information from the senior leaders. This is why you don't want to jump to conclusions the first time you hear something.

Look for Evidence

It is not hard to say the right words. Look for evidence that people know what the words really mean and are living the reality daily. For most of the assessment items, there are objective measures or behaviors that you can observe. For example, ask to see some prototypes. Ask to see some OKRs (objectives and key results). Ask to see the product vision. Ask to see the product strategy. Ask to see some examples of their product roadmaps.

Look Below the Surface

There is no single right way to do product (see the box titled "The One Right Way?" in chapter 9). Anyone who tells you otherwise is just trying to sell you their particular framework or services. If you visit enough strong product companies, you realize there are many effective ways to do product, and if you visit enough weak product companies, you see that there are at least as many ineffective ways. What you mainly care about in an assessment is whether the organization is practicing the product model's principles.

When it comes to how they define roles and responsibilities to divide up the workload, what terms they use to refer to the product concepts, what specific delivery processes they have in place, and what their favorite discovery techniques are, these are all secondary considerations.

Be Kind

Finally, remember that many of the people you interact with will naturally fear they are being judged, especially if they are personally associated with some of the practices that are about to be replaced.

It's important to emphasize that the assessment is not assessing *individuals*; it is assessing the *particular model being used to produce products*.

If handled well, the assessment can help these people start thinking about how they can be a leader of the changes going forward. If handled poorly, these people may consider leaving the company, or, worse, not leave so they can try to sabotage the transformation.

High-Level Assessment

The organizational assessment starts with a high-level overview of how the organization produces products today, then dives into a detailed review of the necessary product model competencies (the different product people and their skills) and the critical product model concepts (what the product people do with those skills).

In the early chapters of the book, we described this holistic view for organizations using the product model. But here you want to look at how the organization currently does this work.

How Products Are Built and Deployed

How often do most teams release? Do they release independently, or do they need to release as a single integrated package?

If a customer encounters a critical problem, what is the mechanism for detecting and correcting it? What is required for the engineers to be able to "release with confidence?"[1]

Who is responsible for ensuring that the new capability works correctly? Is this a largely automated process, or is it largely manual?

What is the level of autonomy in the teams? In other words, how often do product teams complain of having to manage too many dependencies and interactions with too many other product teams to get even simple things done? Are there common complaints about the narrow responsibilities of each product team? Is there a desire for more end-to-end responsibility?

When new features are deployed, are they routinely instrumented so that data is collected? Who is looking at that data?

What is the perception in the organization regarding the speed, quality, and trustworthiness of the engineering organization?

What is the level of technical debt? Keep in mind that everyone has tech debt. But is the situation severe? What are the symptoms? Is there a plan for addressing it? How far along are the teams?

How Problems Are Solved

How is work provided to the product teams? Is it in the form of roadmaps of features and projects? From where do those features and projects originate?

When the team gets to work on something they've been assigned, what is their process? What are the key roles and responsibilities? Who

---

[1] This popular phrase refers to engineers releasing once they are confident that the new capability works as advertised, and does not cause negative unintended consequences (regressions).

determines the detailed requirements, and how is that done? Do they generally have evidence for their decisions or just opinions?

When and how do engineers enter the picture? Are product designers involved? When and how do engineers and designers participate?

What is the role of the stakeholders in coming up with the details of the solution? How is the team ensuring that the solutions meet the needs of stakeholders?

What level of customer interaction is involved? Are potential solutions tested with customers before the decision is made to build? If so, how is that done?

How often are ideas killed or significantly changed? What is the process by which this happens? Do leaders or stakeholders need to approve decisions like this?

What is the definition of success? Is it shipping a feature? Is it shipping that feature on time? Are there measurable outcomes associated with each feature? What happens if the feature ships, but the outcome is not achieved?

What is the perception in the organization of the people on these teams responsible for defining and designing these features? Are they viewed as deeply knowledgeable of the customers and the business? Are they trusted?

How You Decide Which Problems to Solve

Who decides what work will get done?

Does the company have some sort of annual or quarterly planning process? Is the purpose to establish priorities and funding? How are funding decisions made? Who proposes the projects? Does the company fund projects or does it fund product teams—or individuals?

Is there a product vision? Is there a product strategy? If so, are these created at the level of the organization or at the level of individual product teams, or at some level in between?

If the product teams work off a product roadmap, who decides the items on that roadmap? Is the work coming from the sales

194   TRANSFORMED

organization? Is it coming from the stakeholders? Is it coming from the CEO?

What is the nature of typical items on the roadmap? Are they features or projects? Or are they either problems to solve or outcomes to achieve?

And then, how does whoever is deciding what will make the cut and what won't make those decisions? How do they prioritize the work?

Are there desired business outcomes associated with the items of work? Who defines those desired outcomes, and how do they come up with the expectations?

Does the organization track the percentage of items that deliver the hoped-for results?

The Detailed Assessment

Now that you have a broad understanding of how the company produces products, it's time to look at the specific product competencies and concepts to understand just what you have to work with and where the specific gaps are.

As a reminder, these product model competencies and concepts are described in the earlier sections of the book, so here we just discuss how to recognize the competencies and concepts in action.

Product Model Competencies

Before looking at the individual roles, it's useful to get a sense of the overall size of the organization that produces the products. Get at least rough numbers for the total number of people in each of the roles.

This sounds easy, but in many organizations—since titles are not standard—you'll probably need to dig a little to understand what is meant by each role, and then try to figure out what the mix of people and roles is.

For example, many organizations have some people with the title "product manager," but they also have older roles such as product owners, business analysts, program managers, solutions architects, and more.

Yet, if you ask what is meant by the other roles, in many cases those other roles are doing parts of the product management job.

Product Management

The main thing you're looking for here is if the organization understands the difference between a product manager for an empowered product team, versus a product manager for a feature team, versus a product owner for a delivery team.

You can tell this both by talking with the product managers themselves and by talking with those who depend on the product manager, such as stakeholders or engineers.

How do the product managers spend their time? Are they in meetings all day? If so, what type of meetings? How much time are they spending on product discovery?

How well do the product managers really know the customers? How deeply do they immerse themselves in the data? How well do they understand the product's go-to-market? How well do they understand the other dimensions of the business? How well do they understand the industry, competitive landscape, and relevant technology trends?

What type of training have the product managers had? Have they been trained in the full job, or have they been trained only in a role in a delivery process (e.g., Agile product owner training)?

How well respected are the product managers? How well do they collaborate with the designers and engineers? How well do they collaborate with stakeholders? How are these people perceived by the executive team?

Are the product managers receiving coaching on a weekly basis?

Product Design

You're looking for three main things here:

First, does the organization understand what a product designer is and how it is different from the types of designers used by marketing?

Does the organization understand service design, interaction design, visual design, and, in the case of devices, industrial design?

Second, does the organization have enough of these true product designers, or are the designers jumping between several product teams, doing the best they can?

Third, is the design organization set up like an in-house design agency, where product teams make requests of the designers, or are the product designers embedded in the product teams as first-class team members?

Do product managers create wireframes and then hand those to the designer to make them pretty? When are the product designers brought into the process of discovering and designing a solution? Is it after the product manager has largely determined what she wants?

How often are the designers creating prototypes? What types of prototypes? What tools do they use? How are they testing those prototypes?

More generally, is there an experienced design manager in place who knows what to look for in a product designer, and who provides coaching to the product designers on a weekly basis?

Engineering

The engineering organization is normally quite large compared to the product managers and product designers, and you're not trying to assess the full organization and all the various engineering roles here. Your focus is on the more senior engineers who would be serving in the tech lead role on the product teams.

You're looking to see if these people understand the difference between a senior engineer and a tech lead. In particular, you want to make sure these people care as much about what gets built as how it is built.

Remember that it's normal for many engineers to want to focus just on building, but it's essential that you have at least one senior engineer on each product team who is willing and able to participate in product discovery activities.

An excellent sign is when the tech leads have already been visiting real customers.

You can learn a great deal by talking with the first-level engineering managers about which of their engineers are suitable, and whether the engineering leaders understand the importance of engineering in product discovery, and whether they are providing coaching to their tech leads on at least a weekly basis.

Does the organization have a distinction between a senior engineer and a tech lead? Does it have a role that is explicitly responsible to help with the question of what to build and not just how to build? Is the tech lead an individual contributor or a people manager? If a people manager, how many engineers report? How often do engineers visit directly with users and customers? What is the level of interaction between the tech lead and the product manager? What is the proximity of the tech lead to the product manager?

Is the first time the engineers hear of a product idea at sprint planning? What is the role of the engineers in evaluating whether something actually should be built or not?

Do the engineers have "paved paths" or standards for how to accomplish routine tasks, make decisions, and handle specific types of problems?

Are ideas introduced by engineers, especially technical innovations, well received and carefully considered by the product managers?

Are the engineers responsible for the quality of their code, or do they hand off to QA? Do engineers fix their own bugs?

Are any of the engineers outsourced? If so, what percentage? And which specific engineering roles? Are there already plans to bring these roles in-house?

How are the engineers perceived? Are they simply there to build what people request?

Product Leadership

"Product leadership" refers to the managers of product management, product design, and engineering. Virtually every company has people in these roles, but the real question is, what are their specific responsibilities?

You're trying to figure out if the role is set up primarily as a people manager, or if these leaders are deep into the coaching of their people and the creation of the strategic context.

How do they define their responsibilities? Are they responsible for determining the direction? The measures of success?

Do they explicitly include the strategic context (the product vision, product strategy, team topology, and team objectives)? And do they view it as part of their job to evangelize the strategic context? Or do they view themselves primarily as people managers?

Most important, do they view coaching as one of their most important responsibilities? How much time do they spend each week on coaching?

Are these leaders deeply aware of what is going on in their product teams without micromanaging them?

How do the individual contributors perceive their managers? Are they viewed as committed to coaching? Are they deeply knowledgeable of the strategic context?

Product Model Concepts

Now we're ready to look at the product model concepts. Keep in mind that companies almost always have some form of these concepts in place. For example, while companies that have not transformed rarely have empowered product teams, they do often have either feature teams or delivery teams. Your goal here is to understand the most common situation within this company or business unit.

Product Teams

The most fundamental thing to check is if the company has durable teams or if they still have temporary project teams. In other words, are they staffing project teams based on funding a particular project? And after that project is completed, do the people move to other teams?

Assuming the company does have durable product teams, next you're interested in whether the teams are there just to build backlog items (delivery teams), or if they are there to deliver the requested features and projects on a roadmap originating from the stakeholders (feature teams), or if they are given problems to solve and empowered to discover and deliver an effective solution (empowered product teams).

You also want to try to understand if the product teams have all the necessary cross-functional roles. This relates to the product competencies we assessed earlier, but now you're checking to see if each product team has enough people with the necessary competencies.

Next, try to get a sense for the level of ownership that the people on the product teams feel. This is subjective, but when talking with the team members, do they feel a real sense of agency? Are they vested in what they are building? Do they care about the outcomes? If there's an issue, how do they respond?

Finally, what level of access does the team have to customers, to the data, and to the stakeholders? How often are they visiting customers? How often do they include an engineer when they visit a customer? Does the product manager have good access to the various data tools? Does she stay on top of the latest usage data and trends?

Product Strategy

Product strategy is a bit of a tricky item to assess because the term "strategy" is used for so many different things.

Essentially, what you want to understand is how the company determines the product work that needs to be done. Is there a multiyear product vision? Is there a quarterly or annual planning process that results in roadmaps for each team? Or is each product team asked to work with their assigned stakeholders on creating a roadmap? Or do the product leaders create a product strategy that results in the set of problems to be solve?

Keep in mind that every company and every product team has a set of keep-the-lights-on items that just need to be done. These items are not what we're referring to when we are looking at the product strategy. We're looking for a product leader looking holistically across the many product teams and using data to decide the most critical and impactful problems to solve.

Product Discovery

Most companies that have not yet moved to the product model usually don't do any product discovery. If they do, it is usually design rather than discovery.

The key to look for is how many ideas the product team is testing in order to come up with a solution worth building. If the number of ideas being tested is the same as the number of items being built, then this is likely design rather than discovery.

Ideally, what you are looking for is that the product team is testing out a larger number of ideas but building less than half of those ideas in delivery. That tells you that the team is genuinely testing out ideas and discarding the ones that aren't worth pursuing further.

You also want to determine which product risks the product team is considering as they do their product discovery work. Often, it is just feasibility, and maybe usability. You are looking for teams that are also assessing value and viability.

Does the team know how to run quick experiments, both quantitative and qualitative? Does the team understand the techniques to test product ideas responsibly?

Product Delivery

When assessing product delivery, you want to start with how frequently the product team is releasing. Mainly you care that they are releasing no less than once every two weeks. Ideally, they are doing continuous deployment.

You also want to make sure that everything that's released is instrumented, so that you can know if the feature is working as needed.

In addition, you want to determine if the company has monitoring in place to detect if problems occur.

Finally, you want to see if the team has the deployment infrastructure in place in order to run something like an A/B test to determine if a new capability is providing the expected value.

Product Culture

Assessing product culture is subjective but nevertheless very important to do.

First, try to determine the level of formal process the company is following. What is the priority: following the process or following the principles? Ideally, you are looking for teams that understand the principles and use judgment to inform how they work on particular items.

Second, what is the level of trust? Is it mainly top-down, command and control? Or do the leaders try to push most decisions down to the product teams?

Third, is the organization optimizing for predictability, or is it trying to optimize for innovation? Does the company understand the role of engineers in innovation? Are the engineers asked to care as much about what they build as how they build?

Finally, does the company understand the necessary role of failure in innovation? Are people scared to do something that might fail? Also, does the company understand the techniques in order to fail fast and cheap so they don't fail in production?

Innovation Theatre

Many companies long ago lost their ability to innovate.

But before deciding they needed to get serious about transformation, many of these companies tried one or both of two different approaches to try to rekindle growth and innovation.

The first approach involves going shopping for innovation by acquiring companies. The second approach involves setting up some form of corporate innovation lab.

(continued)

(continued)

Acquisitions are a large and important topic, and one that is beyond the scope of this book. However, when it comes to technology-powered products, it's no secret that most acquisitions end up being considered very expensive mistakes—and we're not just talking about the cost of the acquisition. We're referring to the ongoing cost in legacy system integration, technical debt, and unhappy customers.

But the main topic here is the problem with corporate innovation labs. A critical principle is that a product team is responsible for both discovery and delivery, and one of the worst things you can do is break that work across two different teams.

The most common manifestation of this problem is when a corporate innovation lab is responsible for product discovery, and the product teams are responsible only for product delivery.

The reason it's so important that every product team is responsible for both discovery and delivery is that it's absolutely critical that the same people who discover an effective solution be the ones to bring that solution to market. Otherwise, the passion and excitement that the team feels when they engage with customers and the new enabling technology to solve a problem is lost when things are "thrown over the wall" from the team doing discovery to the team responsible for delivery.

Not to mention the problem of creating two classes of product teams—those that innovate and those that don't.

So, while it's understandable why so many companies try out corporate innovation labs, doing so rarely yields the hoped-for results.

CHAPTER

30

Transformation Tactics— Competencies

Establishing the new product model competencies is usually one of the hardest parts of moving to the product model. It is also one of the most sensitive as you are essentially trying to get people to learn new skills and take on additional responsibilities.

But establishing these competencies is usually among the first things to do as, without the skills, the people won't be able to succeed with the product concepts.

Note that many of the techniques in this chapter are described in much more detail in the book *EMPOWERED*.

Product Model Competencies

New Job Definitions

Please remember that just because you may have some titles that sound the same, the product model introduces very different job definitions and responsibilities.

As a reminder, one of the top reasons for failed transformations is ignoring this reality and thinking you can succeed without taking the difficult step of resetting expectations of the people in the roles.

Success begins by first clearly defining the jobs that are required. Then you can start evaluating which of your people are capable of succeeding in those roles. With capable coaching from their manager or another product coach, many can succeed. But if you don't reset the job expectations, you won't see a reset in behaviors or performance.

It's also important to note that there may be job definitions from *outside* of product and engineering that are impacted as well, depending on the org structure. If you find that moving to the product model impacts critical roles elsewhere in the company, consider expanding this work to include those roles.

Job Reset

Occasionally, one or more of the product model competencies requires a true reset in the organization. This is usually the case where the company had been using the same title, such as "product manager" or "product designer," but it had simply retitled previous product owners or graphic designers.

When this is the situation, it's important for the company not just to set a high bar, but also to clearly message the rest of the company that the role has changed very intentionally.

When you redefine this role, it's important to do so consistently. For example, you might temporarily rename all your current product managers "product analysts" or "product specialists." Then, as people interview for the newly defined product manager role, or as they get trained in the new role, you are careful to use only the new "product manager" title if the person has demonstrated that they are capable of succeeding in the new role.

Role Balancing

To get the most value out of your staff, you will need to have the right balance of product managers, product designers, and engineers. It's not unusual that when a company moves to the product model, they are initially out of balance. Role balancing is related to the team topology discussion in chapter 31.

It's common that when moving to the product model, the company ends up needing fewer but much stronger product managers, a larger number of more broadly skilled product designers, and a strong set of engineering tech leads. And the product leaders often have a very steep learning curve because they have much different roles in the product model.

Experienced product teams typically have one product manager, one product designer, and somewhere between two and ten engineers, while a platform team typically has one (quite technical) product manager, and between four and twenty engineers. Those are wide ranges of engineers because there are many different considerations. Just keep in mind it is usually better to have a smaller number of somewhat larger teams than a large number of small teams.

Dealing with Too Few Designers

One common problem is that the assessment shows you have too few product designers. While you are working to hire more, in the interim there are several choices for how to deal with the shortage.

The first option is to triage the product teams and assign a product designer to only those product teams most in need of design help.

The second option is to hire contract product designers to work on a team until a permanent replacement is found. Note that these would be freelancer hires for some number of months, not tied to projects.

The third option is to ask product designers to cover multiple product teams simultaneously. Note that the contribution of the designer drops off dramatically when split between more than two teams.

None of these options is ideal, and none is sustainable, but they can all help you for a few months.

Dealing with Outsourced Engineers

Moving to the product model absolutely requires insourcing your engineers. We mean this very seriously: Just as you wouldn't outsource your CEO, you wouldn't outsource your key engineers.

That said, insourcing can take some real time. The most important person to insource on each product team is the tech lead. That should be done immediately. In fact, you don't have a product team unless you have a tech lead. Once you have that tech lead, that person can coordinate and communicate with the other engineers, even if some of them are still outsourced.

But you will learn that a smaller number of insourced engineers will consistently outperform a much larger number of outsourced engineers. This is why insourcing almost always results in cost savings in addition to dramatically improved levels of innovation.

All that said, while it will take time, the truth is that if your organization isn't serious about insourcing your engineering, it's not serious about transformation to the product model.

Raising Engagement Level of Engineers

When moving to the product model, sometimes the engineers have been treated as mercenaries for so long that the ones who remain are perfectly comfortable in that model, and say they have little desire to participate in activities such as product discovery.

But you need to make sure that at least the tech lead does care about what you are building. The responsibilities regarding product discovery should be explicitly in the tech lead job description.

More generally, the best way to engage engineers is to bring the engineers along when visiting customers. It is remarkable the impact that can have.

Product Managers and Line-of-Business Managers

In certain situations, the product manager needs to partner closely with another person in the company with a similar-sounding job.

For example, suppose you are a product manager for an online banking digital experience, but a different person is product manager for the actual financial bank accounts (savings and checking). Or, suppose you're the product manager for an e-commerce experience, but a different person is a category manager for a class of merchandise, such as electronics. Or, suppose you are the product manager for a digital experience for media and news, but another person is the editor for the content itself.

In each of these cases, the product manager will need to have an especially strong relationship with these line-of-business counterparts. The good news is that there is a great deal of work to do, and it's not hard to split the work up in a way that makes sense. Normally, the product manager on the product team is responsible for the holistic digital or omnichannel experience, and the other person has responsibility for the underlying content or service.

There is, however, one thing you must always watch out for. Sometimes the business counterpart wants to continue to make all the real product decisions, as they likely did in the past. What they ideally want from product managers is that they step back to more of an order-taker product owner role, much like a business analyst. This would essentially be abandoning the move to the product model, as you're right back to stakeholder-driven roadmaps.

That said, sometimes the reason the business counterpart feels this way is because the particular product manager simply doesn't have the ability to do the necessary job. For companies that don't take seriously the new product management competency, this is all too common. In such cases, it actually may make the most sense to consider having the business counterpart take on the product manager role, assuming they are willing and able to be coached in the new skills required for this demanding role.

New Recruiting Practices

One advantage of moving to the product model is that your company becomes much more attractive to the type of candidates you need to hire.

But in addition to new job descriptions, you'll need to be sure to set up suitable interview teams that both know what to look for in a candidate and understand what the candidate is assessing.

Also, it's important to emphasize that if you want to succeed with recruiting the right type of people, the hiring manager is going to need to step up and take ownership of recruiting. HR can help a little bit, but if the hiring manager thinks that HR can recruit for them, she is unfortunately about to learn why they can't.

Assessments and Coaching Plans

Once you have the right people in the right spots, you will want to immediately assess each person against the skills required to succeed in the job. Doing this allows you to identify the gaps and create a coaching plan for each person.

Onboarding Programs

A major part of transformation includes education. It's one thing to read a book or attend a workshop, but it's another thing altogether to learn hands-on what all these techniques are really about.

There are many ways to educate your product teams and your product leaders, but one favorite and scalable way is by creating a product model onboarding program.

Ideally, product teams attend this program together, often including their key stakeholders at relevant moments during the program.

CHAPTER

31

Transformation Tactics— Concepts

Once the new product model competencies have been established, people are ready to start using their skills to work on the product concepts.

It's important to recognize that skills like product strategy and product discovery can always be improved. There are always new, emerging techniques and tools, and good product companies are always working to improve.

So, when you work on transformation tactics, you're not aiming for perfection or even excellence; you're aiming for competence, and then through ongoing coaching, you help your people continue to learn and develop.

Note that many of the techniques in this chapter are described in much more detail in the book *EMPOWERED*.

Product Teams

Team Chemistry

Even if every member of the product team has strong skills, this does not mean that the product team as a whole will work together well. Sometimes, there are personality clashes or other problems. It's important for product leaders to assess the overall chemistry of each team and either coach or shuffle people to get the teams to operate together effectively.

Team Durability

Many companies coming from prior models are used to shuffling people between product teams on an as-needed basis, especially engineers. However, in the product model, while there certainly will be some cases where moving a person temporarily or longer is warranted, such moves are only done judiciously and after careful consideration of unintended consequences—especially given the level of effort product leaders put into coaching the necessary psychological safety and chemistry discussed above.

Review Team Topology

"Team topology" refers to how the product teams are structured—especially what each team is responsible for. You can think of it as: How do you slice up the pie?

Occasionally, a company will not yet have a team topology—meaning it has not yet created durable product teams of any sort. This usually happens with organizations that are still very project based. In this case, transformation to the product model begins with the very foundational move to durable, cross-functional product teams.

However, most organizations have some team topology already in place when they begin a transformation effort.

Sometimes that topology is already pretty reasonable, and the issue is more about how the teams work and interact. But often the topology is more historical, and there are serious symptoms around long lists of dependencies, a resulting lack of autonomy and sense of ownership, and, in many cases, serious morale issues because people feel like small cogs in a giant machine.

And even if the company didn't have these symptoms, the team topology is a function of the longer-term product vision and the ongoing product strategy. So, if the company did not have that product vision, then almost certainly the team topology is not set up to help the company achieve that vision.

Changing a team topology is very disruptive, so it's important not to do this casually or frequently, but the start of a transformation effort is usually a good time.

The key is to make sure, first, that you have your product vision (see "Create Product Vision" later in the chapter) and then that *both* your product and engineering leaders come up with the new topology *together*.

We emphasize this not because we want to sound inclusive. The issue is that the appropriate team topology is essentially a balance of the needs of engineering and architecture, with customers and business objectives. We need to solve for both sets of goals, and one of the most common problems is that just one or the other drives the topology.

There are many considerations in coming up with an effective team topology, but it's also important to keep in mind that no topology is perfect, as they are necessarily based on a set of competing goals and compromises.

One situation to watch out for is that many companies ready to transform have too many small product teams. Often we suggest that the companies adjust to a smaller number of larger-scope teams. It is remarkable how much better the organization can perform, and how much morale can improve, where there are fewer dependencies, more end-to-end responsibility, and, overall, a stronger feeling of sense of ownership and empowerment.

Dealing with Distributed Teams and Remote Employees

When it comes to the geographical location of the various members of your product teams, there are of course many considerations.

What is important for transformation purposes is that while product delivery can work quite well when members are dispersed, it is a much different situation with product discovery. That's because product discovery depends on very close collaboration among product manager, product designer, and tech lead, which is much easier when they are colocated.

Which is to say, when you are deciding whom to put on which product teams, the team will benefit if you can colocate these key roles, even if for only two to three days per week.

Product Delivery

There is a very wide range of activities included in product delivery, and depending on the current state, the list of areas needing investment may be long—including such items as tech debt, implementing CI/CD, DevOps, test and release automation, instrumentation, monitoring, A/B testing infrastructure, and more.

Usually, all it takes is one senior engineer who has already worked this way joining the team to show other team members what is necessary.

Even if a specific team doesn't have an engineer who has already moved to this model, often another engineer on a different team has. In this case, you may ask that engineer to first help their current team become an example for the rest of the organization. Then that engineer will spend the next several months helping other teams through the changes. As you might imagine, such an engineer quickly becomes quite valuable to the organization.

A large amount of work may be involved with architecture, tooling, infrastructure, test and release automation, and more, so it's important to realize that these changes do not all have to be made at once. But these these infrastructure changes are a prerequisite to the product operating model, both in properly taking care of customers, and in holding yourselves accountable to outcomes.

If the level of technical debt is getting in the way of the product delivery work, see the box titled "Managing Technical Debt" in chapter 18.

Product Discovery

Many feature teams consider the work they do to come up with the product backlog as product discovery. And there is a little bit of truth to that. But since feature teams don't specifically tackle value and viability risks, and they rarely actually test their product ideas, other than perhaps performing an occasional usability test, their work would more accurately be considered product definition, not product discovery.

But what's important is to make sure that, going forward, product teams know that they need to address *all* product risks.

This is often a large mental shift for product teams, and especially for product managers. In fact, it's not unusual for some new product managers to resist the responsibility that comes along with being accountable for value and viability. Coaching is critical to help people through this change.

Driving Up Customer Interaction

Many companies that have yet to transform have very little actual interaction with their customers, and correcting that is among the first things you need to fix.

Consider telling each product team with a user-facing experience—customer-facing or employee-facing—that they need to start one-hour product discovery sessions three times per week, every week.

Note that this is not just in "discovery phase," or when there's time. This is the heart of the product manager's and product designer's regular day job.

The teams likely will need the coordination help of one of the user researchers, but they need to get into the rhythm of interviewing or testing their product ideas with real users every week.

For some companies, just clearing away the obstacles and excuses to continuous testing represents a major milestone, and a true turning point in their transformation.

In part X, **Overcoming Objections**, we discuss many of the most common excuses you will hear from people trying to explain why this frequent customer interaction would be too hard. It's not. And it is essential to consistent innovation.

Discovery Sprints

Another very effective way of quickly learning the techniques of product discovery is to hold an intense, one-week *discovery sprint* (aka *design sprint*). This is a great way to build trust within the product team; learn the mindset, principles, and techniques of product discovery; and accomplish something very meaningful in a short period of time.

Again, you can lead these discovery sprints yourself, or you can get a discovery coach to facilitate.

Hack Days

A great way to engage engineers in product discovery, and to ramp up prototyping in general, is with *hack days*. These can be either directed (everyone is working to try to solve the same problem) or undirected (work on anything related to the product vision). More on this under the section titled "Product Culture" later in the chapter.

Product Strategy

The *input* to a product strategy is the product vision and the company objectives, and the *result* of a product strategy is a set of problems to solve, which are then assigned to product teams as team objectives (usually in the form of objectives and key results [OKRs]).

Since the primary purpose of the product vision and product strategy is to share with the product teams and the stakeholders, it's important that the vision and strategy be in a form that is transparent and easily shareable.

It is also important to emphasize that many companies moving to the product model may not yet have a product vision or strategy, but they almost always have some form of company planning cycle that ultimately decides which projects to staff and fund. If it's not already clear to the company leaders, it can be useful to highlight the differences in the techniques. Both start with company objectives, and both finish with prioritized work. But the mechanisms are very different.

Create Product Vision

In the vast majority of transformations, the product organization does not currently have a product vision. It might have an artifact it calls a product vision, but that is usually just a simple mission statement, which does not serve the purpose of a product vision.

Realize that if the company was set up with feature teams, then it is very unlikely that there was any unifying vision at play—as feature teams are all about serving the needs of their particular stakeholders.

It's possible, although unlikely, that one or more of the stakeholders had their own product vision. More likely, they were simply doing their best to run their part of the business. This is why usually the transformation effort is what inspires an organization to create its first product vision.

There are many benefits to creating a compelling and inspiring product vision. The good news is that you put the effort into crafting a strong product vision once, and that vision then provides benefits to the entire company for several years (usually three to ten).

Product vision is also something where experienced product leadership coaches can be a real help, and firms that specialize in helping you craft a product vision also exist.

Create Product Strategy

For the same reasons most companies beginning a transformation don't have a product vision, they rarely have a product strategy. More important, they usually don't have the muscles in place to create ongoing, insight-driven product strategies.

So, it is very common for product leaders to need to put in place the mechanisms to make hard decisions pertaining to focus and to gather insights from data, customer interactions, the industry, and enabling technologies.

This is often the first time the product leaders have been asked to produce this type of work, and those leaders will be judged on the quality of their recommendations in their product strategy. Product leaders who have never before done this work often enlist the help of a product leadership coach for this reason.

Portfolio Management

One very common problem when transforming is that there is a very large set of existing legacy systems and not nearly enough people to care for them properly.

In such situations, we recommend doing a portfolio review. This means creating an inventory of all the various systems and components and then putting each one into one of three states.

For any systems you can possibly *sunset* (stop supporting), you should absolutely do so. You may need to generate some analytics showing the actual usage level of this system, and you may also need to calculate an estimate of the ongoing maintenance cost, but everything you can sunset will reduce both the ongoing load and the number of systems that require replatforming.

Next, likely many systems are important to keep running, but they do not warrant continued investment beyond keeping them running. We call this state *sustaining*. The benefit of sustaining as many systems as possible is to free up most (but not all) of the people who are working on those systems to be able to work on the critical areas of investment.

Finally, the third category is *invest*, which is where you are placing your new bets going forward. These are the systems you want to staff fully.

This portfolio review exercise is usually painful, but it is essential if you want to be able to staff your investment areas appropriately.

Funding Product Teams

If you are coming to the product model from one of the project-based models, it's very possible your organization is used to funding projects, and now you need to change to funding product teams.

We talk about the larger change in chapter 24, "Partnering with Finance," but there is a relatively simple technique for making the necessary change without disrupting much of how your finance organization works.

In the old, project-based model, you would propose a business case for a project, and then finance or the senior executive team would decide if they wanted to fund that project.

You can continue with essentially the same general funding model, but instead of putting a business case together for a project to deliver some output in some number of months, you would put the business case together for a product team to deliver some outcome for some number of *quarters*. Finance usually prefers this because the product team is told to focus on some set of business results, which are much more meaningful to finance than the fact of shipping a feature.

Implement Team Objectives

The result of a strong product strategy is a prioritized set of critical problems to be solved, sometimes characterized as bets. Now you need to assign those problems to the appropriate product teams.

Most companies use the OKR technique for assigning these problems to teams, as that's what the technique was created for. If you've never used this technique, or if you tried to use it but failed because you tried to apply it in the context of feature teams, again, you may want to get some help.

Product leadership coaches can help with team objectives, or specialized OKR coaches can be hired. The key to using an OKR coach is to make sure she has real experience doing objectives in the product model. Many OKR implementations focus on sales and marketing organizations, which have very different considerations.

Product Vision or Strategy Sprints

Product vision sprints and/or *product strategy sprints* are ways for product leaders to dive deep into a facilitated week of product vision and/or product strategy.

The idea is to get the right people together, with the right prep work done beforehand, and then to have a very intense offsite-style collaborative meeting to accomplish in a week what would otherwise potentially take much longer.

These sessions can be facilitated by an experienced product coach or by a specialty firm.

Product Culture

Depending on the results of the organization assessment, the cultural issues are usually the most subjective and the slowest to change. It is not hard to say the right things, but people see what the leaders do, especially under stress.

The key is to realize that there will be mistakes and setbacks. But if you can consistently show visible examples of cultural change, over time, people will come to believe it's real.

This point relates directly to the transformation evangelism discussed in chapter 33.

Many techniques can help build the cultural practices and norms that the product model depends on. Here we discuss some of our favorites.

Hack Days

Getting your engineers and, more broadly, the product teams engaged not just in building solutions, but in coming up with innovative solutions to the problems you are trying to solve, is one of your most important cultural goals.

There are many variations of hack days—directed or undirected, monthly or quarterly, data centered, self-organized or product team based—but the higher-order goal is just to get started and evolve based on the response.

Customer Engagement

How many customer engagements occurred in the past quarter? Has this number grown from the prior quarter? If not, why not?

What were the major insights from those engagements?

How well has the organization continued to increase its level of customer knowledge and share what it has learned?

How much has this knowledge spread throughout the product teams? To the engineers?

Innovation Accounting

How many ideas were tested in the last quarter? Has the number increased significantly from the prior quarter? If not, why not?

What role has new enabling technology played in this innovation?

What role have data insights played in this innovation?

Note that these metrics on ideas and tests can be informative, and help indicate where additional coaching may be required, but they should not be confused with business results.

Culture Retrospectives

There are many forms of retrospectives, but at least every quarter it is a good idea to get some of your most progressive product, design, and engineering leaders together to discuss how they think the organization managed to do its work the preceding quarter, and what they think should be the emphasis for the next quarter.

They would normally look at topics such as:

- The level of process versus the understanding of the principles
- How much was still top-down command and control, and how much was pushed down to the product teams
- The level of experimentation in discovery versus products that failed after delivery
- The level of innovation in the ideas that were explored, and the contribution of the engineers to that innovation
- The degree to which members of product teams understand the strategic context and the product model
- The level of trust between the product teams and the various stakeholders and executives
- The organization's ability to deliver on its high-integrity commitments

32

Transformation Tactics— Adoption

In this chapter, we discuss a different aspect of transforming how you work: adoption. No matter what the technique, sometimes you want to be intentional about how you introduce that technique to the organization.

It's important to point out that these various transformation tactics for adoption are not mutually exclusive. In fact, it's common to blend them based on your circumstances.

Any number of factors can influence your specific choice of adoption tactics:

- The relative importance of the specific new capabilities
- The particular people available to participate
- The availability of product coaches or experienced product leaders
- The window of time available for transformation
- The current demands of the business

- The number of existing commitments
- Dependencies on other teams or systems
- Ongoing architectural changes
- Countless other possible factors

Pilot Teams

The most basic technique is to begin by starting with one or more specific product teams, or a subset of the business unit, with the goal of expanding to the broader organization over time. This method is referred to as *pilot teams* or *pilot organizations.*

Some companies—usually those that have a very strong sense of urgency to transform—attempt to do everything at once.

This can work, but the problem is that there will almost certainly be collateral damage.

An important concept from the product world very much applies to transformations as well.

It's really a question of how fast people can adopt change.

Some people love change (early adopters). Most people don't love it, but once the kinks are worked out, they are fine with the changes (early and late majority). And then there is always a subset of people who hate change no matter what it is (laggards).

This concept is important for your product work because when you deploy changes, not everyone can digest those changes at the same pace.

But this concept plays a major role in transformations as well because, after all, your organization is made up of people, and you are about to make big, disruptive changes to these people.

If you thrust these changes on the organization all at once, then some people will be fine, but most will not be happy.

And it's important to realize this is true largely independent of what the changes are. This behavior plays out even in cases where the people in question have been begging for the changes. They may indeed want the changes, but they want them in a way they can digest.

One effective transformation strategy is to focus on helping the organization absorb the changes using pilot teams.

Pilot teams are product teams (and their associated product leaders and stakeholders) that have volunteered to be on the leading edge of these changes. They want to be the first to try out this new way of working. They understand that they'll be the ones running into the problems, and they'll be the ones who will need to figure out ways around the obstacles.

It's important that we do whatever we can to set these teams up to succeed. For example, we would not want a pilot team that does not have *competent, skilled people covering each of the product model competencies.* Similarly, we would not want to pick a team whose key stakeholders are not also eager to make the change a success.

The main benefit of pilot teams is that the rest of the organization doesn't have to deal with all these changes, especially when the kinks are still being worked out. They can watch from the sidelines and have time to get used to these ideas.

Large organizations (enterprises composed of multiple business units) often do the same thing with business units. One of the business units volunteers to try out the changes, and if things go well, the changes spread to other business units. If things do not go well, the testing and iteration is contained in one business unit unless and until it gets things working.

In general, when using pilot teams, it's better to go deeper with a smaller number of pilot product teams than it is to go shallow with a wider number of teams.

Product Model Dimensions

In large companies, often the immediate needs of different parts of the product organization have different priorities. In these cases, it can be helpful to consider the three different dimensions of the product model:

- Changing how you build
- Changing how you solve problems
- Changing how you decide which problems to solve

When you're balancing different priorities, a second very common transformation strategy is to start by having different groups or teams deploy a *subset* of the product model.

There are any number of ways of defining the various subsets, but one way that is particularly effective is to define the subsets by the three different dimensions:

- One group might just focus on changing how you build (product delivery).

- Another group might focus on changing how you solve problems (product discovery).

- And a third group might focus on changing how you decide which problems to solve (product strategy).

This method allows each of the three groups to focus on the relevant product model concepts.

Working on changing how you build is usually fairly independent of changing how you solve problems and changing how you decide which problems to solve, but the latter two dimensions are interrelated. Product strategy generates problems to solve, and you use product discovery to solve those problems.

But what do you do if you want to work on changing how you solve problems before you tackle changing how you decide which problems to solve?

In this situation, there's a technique that can be used in any organization to convert a conventional product roadmap into problems to solve with desired outcomes. These are called *outcome-based roadmaps* (described in chapter 8), and they are often used as a transition tool as companies move to the product model by starting with changing how they solve problems.

Top Down and Bottom Up

You have probably already noticed that many of the changes apply to product teams, and other changes apply to product leaders.

Here is another good way to naturally divide up the work:

One product coach or leader focuses on developing the skills of the product teams (bottom up), and another focuses on developing the skills of the product leaders and the strategic context (top down).

Coaching Stakeholders

In some companies, a major driving factor is the product/stakeholder interaction, and each side needs to be coached to work effectively with the other.

In this case, you might have one product coach or product leader focus on the product teams and another product coach or product leader focus on the stakeholders.

You seek stakeholders who are very eager to test out the product model and have strong credibility in the organization. Those stakeholders would serve as transformation sponsors. The product team or teams that work closely with that stakeholder would work together as a unit to test out the changes.

The emphasis here is on effective collaboration, where the product teams and the stakeholders learn much more about each other and what they each can bring to the table—letting this serve as an example to the broader organization.

As such, it's helpful to select efforts where an effective solution that solves for all parties is not obvious and will require real give-and-take to identify.

Stakeholder Briefings

A technique related to coaching stakeholders is to hold a separate briefing on the product model just for interested stakeholders where you go deeper into common objections and confusions and give the stakeholders a safe place to raise concerns and work through difficult topics.

Managing Existing Commitments

In most organizations, transformations don't just start with a clean slate. Very likely there exist a number of commitments that still need to be fulfilled.

Certainly, you hope the product leaders will go through the list of existing commitments carefully and creatively to try to reduce the list as much as possible. If there are other ways of satisfying the customer, it probably makes sense to do that. However, likely some commitments remain.

In this case, there are a couple of options.

The most common approach is that some subset of product teams is identified to deliver on these existing commitments before the teams are ready to transform.

The downside of this technique is that for some number of months, you have a potentially large subset of the organization running one way (very project based) and the rest another (product based). And if the existing commitments can't be delivered quickly, then team morale can suffer.

The alternative is that each product team could have some number of preexisting commitments as well as their own new product work.

The advantage to this technique is that it feels more equitable to the various teams. The disadvantage is that it is very hard to do both simultaneously, and likely this choice will delay both delivering those commitments and also the date everyone can get to the new model.

Note that these existing commitments are not to be confused with ongoing keep-the-lights-on work. Every team has those items, both before and after transforming.

CHAPTER

33

Transformation Evangelism

O nce you've done an assessment and have a good understanding of the various tactics you might use to transform the product model, we encourage companies to create a transformation plan. We hope you will write this plan down, and have a very clear owner of making the plan happen.

The Transformation Plan

Ultimately, the transformation plan will represent a good deal of work.

And as with any such enumeration of work, it will be important to assign ownership and accountability to each item of work. There's nothing wrong with having several people working together on any given task, or the entire effort, but in most company cultures, without ownership and accountability, you get little in the way of results.

Now that you have a plan to transform your organization, you need to execute on that plan.

This effort will need to continue for a significant period of time— usually somewhere between six months and two years—and it is very easy for an organization to lose focus or interest. You will need to work hard to avoid that.

Note on Task Ownership

As with most difficult efforts, the key is to have named owners who are responsible and accountable for specific transformation work.

Similarly, it's key to have a named senior product leader who takes real ownership of this overall transformation plan—tracking and reporting progress.

Please don't make the mistake of thinking that a committee is the way to make progress on any large and difficult project. Identify owners, empower them, then hold them accountable.

Continuous Evangelism

Transforming to the product model really does require continuous evangelism.

Product leaders will need to frequently and continuously spend time with product teams, other product leaders, stakeholders, executives, and especially those who have objections or resistance, reminding everyone of the product model, the strategic context (especially the product vision and product strategy), and the areas of progress.

The Value of Quick Wins

There's no question that transformation is a long game. For that reason, it is very helpful for you to have some quick wins.

Here are some good examples of milestones worth evangelizing and celebrating:

- A team that was releasing monthly achieves consistent two-week releases in delivery.
- A team identified, fixed, tested, and deployed a critical fix for a key customer in record time.

- A team decides to kill a popular product idea early in the process based on learnings from an inexpensive experiment.

- A team that has never visited with customers starts doing so and shares its experiences and insights.

- A team uncovers a big insight—positive or negative—from its product discovery work.

- A team that's struggled with dozens of competing priorities now has a single clear problem to solve, and measure of success.

- A team comes up with a truly innovative solution to a tough problem.

- A team achieves an important business result.

- A new compelling product vision has been created and shared with the organization.

- The first insight-driven product strategy is shared.

- A well-respected stakeholder shares positive experiences collaborating with a product team.

Be sure that whenever these kinds of milestones happen, people in the organization immediately notify the owner of the transformation plan and they are added to the monthly update.

Constantly Beat the Drum

The key is a constant drumbeat of transformation evangelism progress.

Be sure you show the organization and its leaders and stakeholders the progress that has been made no less than once a month.

And of course, for that to go well, you'll need to make sure there is real progress to report.

It's hard to emphasize how important it is to keep the drum beating.

As soon as you have a single product team demonstrating real results, make sure every other product team sees those results. Celebrate the accomplishments along the way. Show the other teams what is possible.

Transformation Setbacks

As difficult as it is to successfully transform to the product model, several companies that have managed to succeed and enjoyed remarkable financial rewards have not always been able to hold onto their progress.

These are a few examples to look out for.

The CEO or Key Product Leader Leaves

Eventually the CEO moves on. Of course, the board needs to select a new CEO. But in too many cases, these boards seem to completely forget where they were before they transformed, and they recruit a new CEO who has absolutely no idea how to run a product company. That person proceeds to destroy literally years of work in just a matter of months.

Or, a key product leader moves on, and the credibility of the new model was tied to that person and not to the model. When a new person comes in, they don't inherit the credibility and regression can occur.

Usually, by the time the product organization realizes what has happened, the damage has already been done.

This illustrates why it is so important to keep your board of directors *fully* informed on the product model transformation and make sure they realize that the financial rewards the company has enjoyed are a direct result of the transformation.

Scaling Across Company

Once the initial business units have transformed, the company may decide to transform the remaining business units. Each of these transformations is hard and sometimes, as time goes on, the whole model gets watered down because the people leading the transformation are just going through the motions.

The main thing to keep in mind here is to do a fresh assessment with each new business unit. While there will be similarities, it is the differences that are important to understand. Also realize that the level of motivation may be different. Usually, the motivation is highest with the first business unit, which is why it wanted to be the first. But in any case, remember the critical role of the leader of the business unit.

The PMO Returns

Even in organizations that are thriving in the product model, often executives still miss the days of command and control. If one of those executives gets promoted, many times the first thing they do is resurrect the old-style PMO, and slowly but surely the old culture creeps back in.

If you've ever seen how hard Amazon works to make sure it continues to behave as a "Day 1" company and take every precaution to avoid becoming a "Day 2" company, this is what Amazon is worried about.

Business Climate Changes

During challenging business environments, some leaders may want to grab the steering wheel back. Or the opposite situation: If a life-or-death business event motivated the successful transformation, once that threat has passed and the business stabilizes, there may be some desire to get everything back under control.

During the pandemic, a number of companies showed what they were capable of and achieved remarkable results. But afterward, without the pressure of a crisis, companies had a harder time keeping people from reverting back to old behaviors.

Likely it will be up to the CEO or business unit leader to keep everyone focused on the principles of the product model and not regress to the comfort of the past.

CHAPTER

34

Transformation Help

Please note: We realize that what follows runs the very real risk of sounding self-serving, as at SVPG we are essentially product coaches as well. However, we are literally only a few people, and odds are we don't have the bandwidth to help you even if you wanted us to. Mostly what we do is refer companies to product coaches whom we know and trust. We don't charge companies anything for these referrals, and we don't accept kickbacks from any product coaches whom we refer. We simply want you to be able to trust that when we refer someone to you, they know what they're doing.

How does a company that has never worked in a new way learn that new way of working?

The most important thing to realize is that your company depends on building products that your customers want to buy. And your products are created by your product leaders and product teams.

To transform, your product leaders and the people on your product teams need to learn new ways of working.

While training and books can help (if the trainers and authors know what they're talking about), it is never sufficient.

In this chapter, we discuss the different approaches to getting help, and we also share with you the profiles of several very strong product coaches so that you will know the type of person to look for.

In the best product companies, the way product managers, product designers, and engineers learn their craft is from their managers.

Managers as Coaches

The legendary coach Bill Campbell was famous for saying *"Coaching is no longer a specialty; you cannot be a good manager without being a good coach."*

This is why, at the best product model companies, the number-one attribute of strong managers is that they are considered good coaches.

But what to do when your managers have never worked this way?

One solution is to hire senior product leaders (especially leaders of product management, product design, and engineering) who have already worked in this model, and then they can lead by example. This method is very effective and is one of the most common paths to a successful transformation. (This is what happened in the two transformation stories you've read so far.)

However, often this is not enough for two reasons:

First, in many cases, the company has some number of senior leaders who have never worked this way before.

Second, even if the leaders do have the experience, they simply may not have the time to lead the organization as well as personally coach all the people who need coaching.

In these cases, an external product coach can be the difference between success and failure.

In-House Product Coaches

For larger companies, it may make sense to hire one or more in-house product coaches.

Normally, product coaches are temporary—there to help you with a specific product team, or set of product teams, or for the duration of your transformation. But some companies like to retain one or more product coaching staff indefinitely.

Remember that the best solution for ongoing coaching is staffing these management positions with people who can effectively coach your individual contributor product managers, product designers, and engineers.

But even then, it can make sense to have one or more specialists available to focus on sharing best practices and helping with training and onboarding of new team members.

Just be very careful here to ensure that only skilled professionals are in this role. Again, this should be obvious, but it is remarkable how many companies put people in these in-house coaching roles who don't have necessary skills and instead just end up institutionalizing bad practices.

External Product Coaches

There are many types of coaches out there, but when it comes to moving to the product model, there are four common types you should know about.

Delivery Coaches

If your company is not yet able to do frequent, small, reliable, and uncoupled releases—no less frequently than once every two weeks—then you have some serious work to do.

Delivery coaches are experienced engineering professionals who help product teams through the hard work of getting the test and release automation, instrumentation, monitoring and reporting, and deployment infrastructure to the level it needs to be.

In certain cases, Agile coaches who have the actual work experience can help with this delivery coaching work. But Agile coaches who just focus on the delivery process (e.g., Scrum, Kanban) are not going to be especially helpful here.

Discovery Coaches

The foundation of product coaching is discovery coaching. These people coach product teams on the techniques of product discovery.

We see more discovery coaching opportunities out there than anything else, simply because so many product teams are asking for help. Moving from a feature team to an empowered product team primarily means learning how to discover a solution worth building, and this is what product discovery is all about.

All successful discovery coaches we know are former product managers, product designers, or tech leads who have learned the skills and techniques of effective product discovery and love to share their knowledge with others.

Product Leadership Coaches

As you read earlier when we discussed product model competencies and product model concepts, product leadership is hard. This is especially the case when trying to move to the product model. There are now big and critical topics like product vision, team topology, product strategy, team objectives, and, of course, developing their own staffing and coaching skills.

Many people, especially in rapidly growing companies, have had "battlefield promotions," and now find themselves leading product, design, or engineering, but they know they need help.

All the successful product leadership coaches we know are former heads of product, heads of design, or heads of technology who have figured out how to tackle these big topics and who want to share what they've learned with others.

Transformation Coaches

The final type of coaching we see are those who help guide the senior leadership team through the necessary changes in mindset and culture needed to move to the product model.

Transformation coaches usually work directly with one of the senior leaders of the company.

Senior leaders know they need to change product and engineering, but they know it will be even more difficult to change the way they fund (finance), the way they staff (HR), the way they sell and market (sales and marketing), and more.

What makes this type of coaching especially tough is that the vast majority of CEOs simply won't trust the future of their company to a product coach who has not "been there and done that" as a senior leader with other large and complex companies.

They need someone who can hold their own with the company's CFO and explain to the head of sales why changes are necessary yet can also engage directly with engineers, designers, and product managers.

In every successful transformation we know, there was someone who knew what was necessary and had a trust-based relationship with the CEO.

Finding a Product Coach

One way or another, it's essential that you have people who can coach those who are trying to learn the new way of working.

Which leads to another of the most common sources of failed transformations: hiring the wrong people to help.

Most companies know they need help, but they don't know enough about the new way of working to judge the people they hire to help them.

As obvious as this sounds, you need to make sure you have product coaches who actually can help your people learn the skills they need to learn. What is less obvious is that most of the people who try to sell you their services literally do not have this experience.

To be explicit, if you are hiring a delivery coach, they need to have a track record of helping product teams get their test and release automation to the point where they can do continuous integration and continuous delivery.

If you are hiring a discovery coach, they need to demonstrate to you that they have real experience with product discovery as either a product manager or product designer at a serious product company.

If you are hiring a product leadership coach, they must demonstrate to you that they have leadership experience with product vision, product strategy, team topology, and team objectives at a serious product company.

If you are hiring a transformation coach, they must demonstrate to you that they have the experience leading a company through a successful transformation to the product model.

Why emphasize this? Because there exists a whole coaching industry that comes not from product companies, but instead from either specific software processes or management consultancies.

But this is too important to leave at the theoretical level. So, we want to share with you the profiles of seven product coaches we know personally.

It's important to emphasize that we have no financial relationships with any of these product coaches. But we often recommend them, and other product coaches like them, because we believe they have the knowledge and skills to help companies succeed.

Product Coach: Gabrielle Bufrem

Path to Product

Gabrielle is truly a citizen of the world. She was born and raised in Brazil, then went to study in the United States. She speaks four languages fluently. She has lived and worked in ten different countries spanning Europe, North and South America, and Asia, and she has built products across nine major industries.

Gabrielle fell in love with product during her internship at Google. The internship was in marketing, but she asked so many product-related questions that her manager said she sounded like a born product manager and that she really should explore that role. Gabrielle was immediately taken with the role and adjusted her final course load at Brown University to incorporate computer science and design.

After her studies, she joined Education First in Boston and then Switzerland as their first product manager, then spent five years—first as a product manager, and then as a product leader—at Pivotal Labs, mostly in Silicon Valley, but also working in the Paris and Singapore offices.

As she quickly moved up the product leadership ladder at Pivotal, Gabrielle focused increasingly on her coaching responsibilities to her team. She established a strong reputation as a product leader who knows how to develop strong product people and is willing to put in the time and effort to do so.

After Pivotal, Gabrielle joined Little Otter Health, a mission-based startup in the mental health field, as their first head of product and design, then she proceeded to build a modern product team as well as significantly improving the product offering.

Journey to Coaching

During her time at Pivotal, and then Little Otter, Gabrielle realized how much she enjoyed helping others, and she began teaching product management, speaking at industry conferences, and unofficially coaching people beyond her direct reports. She had incredible mentors and managers who shaped her view on product and on what was possible for her career.

Gabrielle soon realized that what she loved most and did best in her previous roles was coaching and developing the *people* who led these product organizations. This realization inspired her to move into product coaching full time.

While Gabrielle's specialty is coaching and developing product *leaders*, her superpower is quickly establishing a strong personal relationship and building trust. Trust is what allows Gabrielle to deliver the frank, honest, and critical feedback that helps people truly break past old limits to reach their full potential.

Gabrielle is known both for her kindness and her toughness. She is able to be truly candid because the people she coaches know that she genuinely cares and is there explicitly to help them succeed.

Much of this derives from her personality and life experiences. People can relate to her easily, and they see that she understands both product and people, and genuinely wants to help them meet their goals.

Product Coach: Hope Gurion

Path to Product

Hope began working in product in the early days of the internet, when she was in the middle of the mix helping to create many popular early consumer services, including online shopping, real estate, and job and career services.

Hope's broad knowledge of business—especially business development, sales, marketing, advertising, and finance—combined with her passion for creating products and services that customers loved, put her on a proven path to product.

Eventually, this led to building and leading the product organizations at major brands, including CareerBuilder and Beachbody, where she built a reputation for building strong product teams that were skilled at solving problems for customers in ways that met the needs of the business.

In many of these companies, building the product organization meant moving beyond stakeholder-driven feature teams and transforming the broader organization to the product model.

After doing this multiple times, Hope realized that she possessed unusual and valuable experience, which is especially important to product leaders who have the responsibility but have not yet experienced this way of working.

At the same time, Hope witnessed so many companies that didn't know how to recruit or attract an experienced product leader skilled in the product model. These companies often ended up placing a high-potential leader from an adjacent function—such as engineering, design, marketing, or business strategy—into the product leadership role, but without any mechanism to set up the new leader for success.

Journey to Coaching

In 2018, Hope made the decision to become a full-time product leadership coach. She also developed the *Fearless Product Leadership* podcast to help aspiring product leaders around the world learn what it takes to be a strong product leader.

In addition to her work coaching product leaders, Hope coaches product teams in continuous product discovery through her work with Teresa Torres, author of *Continuous Discovery Habits*.

One of the themes of Hope's career has been working at the intersection of business and product. Hope is especially interested in helping leaders and teams understand the dynamics of their business and precisely how their product work drives business results. Says Hope: *"During my time leading product teams, nothing thrilled me more than witnessing my teams reach their goals. Now, as a coach for product leaders, I'm thrilled when I witness their transformative shift in perspective. When they grasp the heart of a persistent challenge, it ignites newfound confidence and clarity, empowering them to fearlessly confront it head-on."*

Product Coach: Margaret Hollendoner

Path to Product

Margaret discovered product management almost by accident after years of trying to blend her love of problem-solving and learning with her strengths in teams, people, and communication.

Upon graduating from the mechanical engineering program at Stanford, she took a role as an applications engineer at a controls company in order to be customer facing. Margaret loved being on-site with customers, especially when teaching operators how to work machines and diagnose issues.

Margaret initially thought she wanted to pursue a doctorate in thermodynamics, but when she realized how many hours of isolated research were involved, she left Stanford to seek out a more people-facing role.

She decided to apply to Google, where her interviewer (to whom she is forever grateful) observed Margaret's mix of skills and motivations and guided her into product management.

Over the next 18 years, Margaret learned product management from within Google. She soon learned the value of being "the person who crossed the street"—going between the engineering and customer-facing buildings to ensure product strategy and product discovery was close to the needs of users.

It wasn't long before Margaret was managing teams of product managers delivering on entire product lines.

During her nearly two decades at Google, she had the opportunity to work on all types of products, and with all manner of teams—from consumer to B2B, across health, video, commerce, and ads. She led organizations of 100+ designers, engineers, marketers, and more, and managed teams of product managers to discover, develop, iterate, launch, and land successful products such as AdSense, YouTube Video Measurement, and Google Fit.

Margaret also worked on her fair share of discovery efforts that never saw the light of day, such as brick-and-mortar store offers, terrestrial TV measurement, medication management, and more. To build and lead motivated, fast-paced, and effective teams, she learned to celebrate success as well as failure and nurture both within the culture.

Journey to Coaching

Margaret began her career at Google with a couple of years at the headquarters in Mountain View, but then she returned to her native England where she spent 15 years as a product and tech leader in Google's London offices.

This position involved building and coaching innovative and diverse teams, often from scratch, that could successfully execute on technology-powered products.

It was during this time that Margaret truly began to focus on her interest in coaching and developing others. She hired, developed, mentored, and advised across tech functions, including product, engineering, design, and research.

Margaret built peer-mentoring communities across the women PMs, London PMs, and managers of PMs. She was a founding member of Google's tech apprenticeship program, bringing in recruits to be developed into software engineers. She developed and delivered training and brown-bag talks and supported individuals converting to product management from other roles.

Margaret discovered that her passion for motivating and developing individuals to do incredible things matched her enthusiasm for building product vision and strategy herself. Developing product people was an amazing combination of both.

Margaret was so engaged with the organizational development aspects that she took a temporary, six-month assignment into the tech HR team to help Google improve how cross-functional tech teams could collaborate and thrive.

Outside of Google, she began mentoring and advising startups and individuals across the tech landscape, eventually giving her the motivation to become a full-time product coach.

This allowed her to focus on the parts of the product leadership role she loved the most: working with individuals and organizations to support them in crafting their product vision and strategy, align across stakeholders, and develop the most effective cross-functional product teams.

As a product coach, Margaret is energized by the passion and dedication she encounters in individuals across all types of companies and industries, and she's inspired by their commitment to whatever their mission may be, whether it's solving women's health crises or selling parts to factories.

It is these people who motivate her to continue coaching, and it's a unique opportunity to partner with leaders across industries to develop thriving, successful, cross-functional product teams that can have the greatest of impacts through technology.

Product Coach: Stacey Langer

Path to Product

Stacey began her professional career by joining an electronics retailer, Best Buy, just as the company was starting to expand from physical retail to online.

Stacey's career grew with the company as she took on progressively more responsibility and had the opportunity to learn firsthand many of the product roles as they were introduced to Best Buy.

Over two decades at Best Buy, Stacey worked in content creation, product design, product management, and testing and optimization—initially as an individual contributor, then as a manager and director of product, and eventually as the company's senior vice president of product.

Along the way, she witnessed the many challenges up close, and learned firsthand why an outsourced, waterfall technology model would not be able to meet the needs of customers or the business.

Over time, Best Buy determined that it wasn't enough to have an e-commerce site. It needed to transform to the product model to better meet customer needs across channels.

Stacey was one of the early managers asked to lead this effort. As such, she had to tackle the challenges of moving from outsourced IT to in-house product teams, and she had to staff and coach for all the key product competencies—including product management, product design, and engineering.

The fact that Best Buy still exists and thrives, while so many other electronic retailers are no longer part of the competitive landscape, is testament to the company's ability to adapt and change and continue to provide real value to its customers.

With her success guiding a major retailer through the challenges of transformation, Stacey decided to help do the same with the United

States Digital Service, ultimately serving as the digital services director and deputy CTO of the US Department of Veterans Affairs to apply what she had learned to help millions of veterans better discover, apply for, track, and manage the benefits they earned.

Journey to Coaching

One of the most fulfilling parts of Stacey's career has been creating space for talented teams to build great experiences for their users. She believes this is most possible within the product model, where a team of people with different capabilities, approaches, and perspectives all work together to create a solution that none of them would have come up with on their own.

After working in product for many years, and leading transformations at very large enterprises, Stacey found herself being asked more and more to help leaders in other organizations who were taking on the challenge of transforming their teams.

Stacey came to realize that the work of coaching leaders and teams was both impactful and something she loved to do. As a product coach, Stacey applies her real-world product leadership experience to help leaders and teams move to the product model, centered on their users, and bringing the best of themselves to create meaningful products.

Product Coach: Dr. Marily Nika

Path to Product

Marily grew up in Greece, and she has a unique and inspiring story that demonstrates how passion for technology and perseverance led her to a successful Silicon Valley career.

When Marily was seven years old, she discovered her brother's old BASIC programming book, and she spent countless hours learning on the family's Amstrad early personal computer. Coding became Marily's passion, and she knew from this young age that she wanted to study computer science.

Unfortunately, Greece's unique academic system didn't allow students to directly choose their area of study, and Marily was instead enrolled to study economics. Almost everyone around her tried to convince Marily to let go of programming, but her parents and her first mentor told her that it matters less what you study in school and more where your passion is. Sure enough, Marily ended up getting a scholarship from Google, and eventually a doctorate in machine learning from the prestigious Imperial College in London.

When Marily joined Google in 2013, she had no idea what product management was. But she was soon working with product managers and became mesmerized by the job.

Thankfully, Google offered an internal product management rotation program, so she was able to see if the product role was a good fit for her. It wasn't easy to transition from her previous technical role to product, but Marily never looked back.

Marily spent eight years at Google—both in London and in Silicon Valley—building AI products for Google Assistant and AR/VR, and then two years working at Meta's Reality Labs.

Journey to Coaching

Given Marily's success with AI products, she naturally became a magnet for people asking for mentoring and career coaching—first from colleagues inside her companies, and soon beyond.

Marily came to realize that not only was she good at mentoring and coaching, but she truly loved every minute of it. And her coaching naturally expanded to writing and teaching.

Today, Marily is a popular product coach and speaker at major conferences, and she teaches AI product management through her own online courses and through Harvard Business School. More generally, Marily works tirelessly to help as many people as she can pursue their passions in AI technology and product.

Product Coach: Phyl Terry

Path to Product

Phyl has pursued a somewhat unconventional but extremely effective path to product.

After being part of the first startup Amazon bought, and after a stint at McKinsey, Phyl joined Creative Good (CG) as CEO and proceeded to build and run it for 15 years.

Creative Good built a reputation as a premier partner working with many of the top product teams and product companies in the world to help them get closer to their customers through hands-on user research, which the company termed "listening labs."

During this time, Phyl and the CG team engaged with many of the industry's leading companies, including American Express, Apple, BBC, Facebook, Google, Microsoft, Nike, and *The New York Times* as they all worked to gain a deeper understanding of their users and customers.

To succeed at this work, Phyl engaged personally—not only with the product leaders and product teams, but also with the senior executive teams, who sometimes participated only reluctantly.

All of these experiences gave Phyl an unparalleled opportunity to get to know strong product teams and product leaders from across the industry.

Journey to Coaching

After the dot-com bubble burst, Phyl decided magic could happen when product leaders connect with one another. So, Phyl saw the opportunity to combine the two things they loved most: coaching product leaders and connecting product leaders into a peer-network community.

So, in 2003, Phyl founded Collaborative Gain and has been building this community for the past 20 years. It is the largest such network in the world, with several hundred product leaders from product companies around the world.

Phyl has helped countless product leaders not only to become more effective at their jobs, but also to find more satisfaction and happiness in their careers and lives.

They have learned that good coaching goes well beyond good listening. It obviously requires knowledge of the substance. It requires a continuous-learning mindset. It requires a real willingness to roll up your sleeves and do the work—both the coach and the coachee. It requires a deep understanding of company politics and culture. It requires a willingness to have difficult and even uncomfortable interactions. But most of all, it requires real *thinking*—thinking hard as you formulate your coaching advice and coaching the coachee in how to think through hard problems.

Phyl also loves to coach through writing, and they have published *Customers Included* and *Never Search Alone: The Job Seeker's Playbook*. For the latter, Phyl has built a free, volunteer-driven global community of thousands of product managers (and others) looking to grow their careers and find great roles.

Product Coach: Petra Wille

Path to Product

Petra's journey to product began more than two decades ago. She began her career working in engineering for several years, but from early on, she was fascinated by the creative process of building products.

Driven by a desire to understand the big picture and to make a greater impact, she transitioned from engineering to product management. As a product manager, she was able to get involved with every aspect of product creation—from concept to launch.

Petra worked for several high-profile German companies, and these years of hands-on experience helped her learn the ropes of product management and product teams, and understand the intricacies and nuances of creating great products.

Soon, Petra was promoted into leadership positions, and she served as a product leader for two major German tech-powered companies: XING and tolingo. In her new role, she had her first real responsibility for coaching, and she immediately understood the importance of people development in her work. In addition to the joy she found in helping others, she discovered that her focus on coaching had the additional benefit of attracting and retaining talent.

Journey to Coaching

Petra's transition into full-time coaching felt like a natural next step. When Petra left her head of product role to go out on her own, she had the freedom to lean into her coaching skills. She began conducting workshops and coaching sessions, gradually developing a toolkit of her own personal coaching strategies that she has continued to build and develop.

As a product coach, Petra has empowered many product managers and product leaders to find their own pathways to success. She's witnessed people going from feeling lost amid organizational changes, to confidently steering their teams toward success.

As with all good product coaches, her approach doesn't involve forcing a one-size-fits-all strategy or framework onto the people and organizations she works with. Instead, she tailors her coaching to the individual needs and context of the people and companies she is trying to help.

Petra's coaching journey has led her across a broad spectrum of industries, from healthcare technology to e-commerce to container shipping. But the common thread is helping people create successful products and fulfilling careers. As Petra explains: *"A strong product coach meets people exactly where they stand, regardless of how early or advanced they are in their product journey. This ability comes from a foundation rooted in real-world experiences in product, providing a bedrock of understanding from which they can offer relevant and actionable coaching and advice. It's this hands-on experience, coupled with a genuine passion for helping others, that allows a coach to bring about transformative change."*

In 2020, Petra published *STRONG Product People: A Complete Guide to Developing Great Product Managers,* and she is also an active leader of the European product community—both speaking at and organizing popular product conferences.

35

Innovation Story: Datasite

Marty's Note: Another very common misconception in the product community is that innovation only happens in consumer products, but that it's a different game when building products for large companies. It is unfortunately true that many enterprise software companies have been slow to move to the product model, and it's also true that sales-driven organizations can be difficult to change. But as this example shows, for the ones that do, the rewards are substantial.

Company Background

You learned in part VII about the Datasite transformation story. But the reason companies invest in moving to the product model is so the organization is capable of consistent innovation. This innovation story shares one such example.

One of the keys to a successful mergers and acquisitions (M&A) business is privacy. During the M&A process, numerous confidential documents are exchanged between the involved parties. If these documents are not handled securely, they can be lost, misplaced, or stolen.

Document redaction is the process of selectively removing or obscuring confidential data from a document to protect sensitive information from unauthorized access or disclosure while still allowing the remaining content to be shared or published.

Redaction typically is applied to text, images, or other media within a document. The process involves ensuring that this content cannot be revealed to the public or, in some cases, sold on the dark web.

Document redaction can be performed manually, where either lawyers or paralegals carefully review the document and apply redaction marks or cover sensitive information using physical or digital methods. Manual redaction requires attention to detail to ensure the selected information is hidden effectively.

In 2019, redacted documents played a prominent role in a major legal and political news story.

One of the lawyers inadvertently filed a poorly redacted document to the court, and the public was able to view previously hidden information.

Due to the ensuing headlines, the case proved to be a wake-up call to the business community, which realized that the tools they were using for data redaction were not as effective as they had once believed.

The Problem to Solve

Datasite knew that to maintain client confidence, it had to build a redaction tool that could ensure the integrity of the documents shared during the M&A process.

The fundamental problem was that while most redaction tools were able to remove or mask a text item or image, they did not destroy the metadata behind it, and a determined person could forensically recreate the redacted data.

Datasite decided there was no adequate solution on the market, so they needed to build a new and truly effective redaction product themselves.

Before the company's transformation, this wasn't something the organization would have been in a position to pursue. But Datasite had been building the skills for this since moving to the product model.

As the product team spent time with customers to understand their redaction needs, the team began to evaluate the various technical approaches that might be used to do this deep redaction.

They quickly recognized that it had to be able to pinpoint select data from hundreds, sometimes thousands, of documents, then remove not only the data but also its metadata in such a way that it did not harm the rest of the document.

Discovering the Solution

The engineers needed a way to cleanse a document yet recreate the redacted data with the right permissions and access to the Datasite system.

As is so often the case, one of the engineers figured out a novel approach to this problem, inspired by an approach used in Google Maps. To prove technical feasibility, the engineers built a feasibility prototype, and demonstrated it to the product team and product leaders.

Next, the engineers needed to prototype and test the customer experience on the Datasite platform.

The team worked with 12 customer discovery partners in Europe and North America to discover and test this proposed solution. This extensive validation testing enabled sales, marketing, and service teams to participate early in discovery with clients in London, Paris, and New York.

The Results

The end result was the discovery of a now-patented technology that automatically redacts a document, or an entire virtual data room of documents, by removing both data and metadata without any possibility of identifying what had been removed. Equally important, documents can also be automatically unredacted when necessary.

Datasite's redaction tool became a significant market differentiator that provided real value and peace of mind to the company's clients.

IX

Transformation Story: Adobe

By Lea Hickman

Marty's Note: What follows is one of the most impressive transformations I've ever seen. Adobe was always a solid company and it created several successful products, but it wasn't operating in the product model, and it knew it needed to change in order to survive the challenges headed its way. In this case, the company's innovation story is what drove its transformation story. The driver for the transformation was the need to create the new Creative Cloud offering. The story is told by Lea Hickman, who was appointed head of product management for the new Adobe Creative Cloud. Lea teamed up with CTO Kevin Lynch, among others, and with the help and support of some very strong senior executives, they were able to pull off what is considered the industry's most financially successful transformation to date.

Many people think of Adobe as a post–internet era company but, in fact, the company was founded in 1982 and has played a key role through many generations of technology, starting with printers and personal computers. Today, the company spans many different businesses, mainly focused on providing designers with the tools and technology they need to build our digital world.

The Motivation

In some companies, there are early indicators that point to the need to change the way the company produces its products.

Sometimes, it is a new disruptive technology, or a major shift in market dynamics. Much of the time, the organizations are able to adjust and react accordingly.

Back when I worked at Adobe, from 2007 to 2014, we were going through just such a time. We were facing major shifts in technology, market, and competitive landscapes.

At the time, I had product management responsibility for the design, web, and interactive tools in the digital media organization. This portfolio included tools like Adobe InDesign, Adobe Illustrator, Adobe Dreamweaver, and Adobe Flash Authoring, among others. We sold our products as individual tools as well as bundles that we called the Creative Suites.

Each Creative Suite had many different bundles based on customer personas. For example, there was a Design Suite, a Web Suite, and others. There was also a suite called the Master Collection, which included one of every tool we sold.

Our primary business model at the time was to sell upgrades of these tools and suites to existing customers, and to sell full versions to new customers.

Each time we would plan for a new release, we would project how many upgrades and full versions of the product we would need to sell to hit our revenue targets. We would project this based on our confidence on whether the upcoming release would achieve product-market fit.

In the midst of our analysis for an upcoming release, the data indicated that what our customers already had might be good enough for them, meaning that the likelihood of them upgrading was not high.

At the time, all the products were quite mature, having been in the market for well over a decade. With the prices continuing to rise, the question was whether we were building enough value into the release to compel customers to upgrade or buy.

Another challenge we had was our long product cycles. When I first joined Adobe, our product cycles were roughly 18 to 24 months between major releases. That changed somewhat over time as we started delivering point releases in between the major releases. Doing this helped us to introduce new features that leveraged new technologies in an interim cycle. While this helped from a revenue recognition perspective, it didn't help with addressing new competition as competitors had already adopted the practice of continuous releases.

As more software was being delivered as a service during this time frame, new competitors entered the market. Competitors such as Sketch and Aviary were being mentioned by customers increasingly often. While these competitors were very new to the market, their value proposition of a simple, modern design tool resonated with customers who were using only a fraction of the capabilities of our old, complex tools.

Also around the time, new technologies were having a significant impact on how customers worked. In particular, the iPhone had been out for a few years, and more people were talking about cloud-based services.

The idea that the cloud was going to be where users (designers) stored their data, versus their desktop or laptop, was very powerful because it would unshackle users from specific devices and allow them to access their data and files from any device, anywhere.

This of course would lead to a change in the way designers ideated and designed, and their workflows would change dramatically as a result. It would mean, for example, that more ideas would be captured by the cameras on smartphones, and this would open yet more possibilities.

Our solution for mobile app authoring at the time was very proprietary. We evangelized using Flash authoring and Flex as a way of building rich applications that would run on both Android and iOS, assuming another one of our products, Adobe Air, would serve as a container for those applications. At the time, the idea of writing applications once and being able to run them across multiple mobile operating systems was very compelling.

Although a segment of our customer base supported that, a large portion believed that a native app would perform better, and others felt a standards-based framework would be a better approach.

Those changing workflows meant we needed to rethink not only how our customers built applications, but we also needed to think about how *we* built mobile applications.

This philosophical disconnect created a lot of churn within the company. How could we build native applications ourselves when we recommended that our customers build using our proprietary frameworks?

Another dramatic shift during this time was a change in who our customers were. There was a shift underway from print to digital mediums, and there was a shift from professionals to more part-time hobbyists or aficionados.

While a portion of our sales always consisted of those markets, the trend was becoming clear: New competitors that had simpler tools were more compelling to nonprofessional users.

There was also a change underway in buyer behavior. At one time, Creative Suite buyers would go into their local tech store and purchase the Creative Suite right off the shelf. They would then install DVDs of the software on their laptop or desktop.

Many years earlier, we introduced the ability for users to download the software, and now downloads were becoming the preferred method. With download speeds increasing, it became much easier for people to purchase and download right from their home and office—disintermediating the retailer or reseller, or at least reducing the number of units sold through that channel.

The final area of motivation was in the form of an open letter that Steve Jobs personally published on April 29, 2010, about Flash and why Apple made the decision that the technology would not be allowed on Apple's iOS devices.

This was a substantial blow. Our Creative Professional products had always supported Apple's hardware because Creative Professional customers loved Apple products.

We knew we had to take a different approach.

But knowing changes are needed is not the same as being willing and able to change.

So far all we had was the stick.

We knew that if we didn't make major changes both to how we worked and to our product offerings, we would lose our market position.

We knew that we also needed a carrot.

If we were to make the necessary changes, what would we be able to do for our customers and for our business that we could not do today?

Adobe had a long history and culture around an existing and thus far successful way of working. And while the forces going forward were clearly troubling, it's important to recognize that, at the time, Adobe was the clear leader in the industry—a publicly traded company—with the Creative Suite representing more than $2 billion in annual revenue for the company.

To answer this question, three of the senior leaders at Adobe— our then CTO, Kevin Lynch, the SVP of our Platform Business, David Wadhwani, and the SVP of our Creative Business, Johnny Loiacono— called an offsite. They asked each product category to come to the offsite prepared to present and demo new product ideas that were beyond their desktop tool offerings. Specifically, they were looking for mobile apps and hosted services.

While we knew that hosted services and mobile apps were critical to future value for our customers, we didn't know how far the teams had progressed in their thinking. At the offsite, it became apparent that each area of the organization had a different vision for what it thought customers wanted, and the visions were not aligned.

At that offsite, senior leaders realized that not only did we need a unifying vision beyond just the capabilities that the cloud, mobile, and desktop could bring customers, but we needed a product leader to drive that vision. The leaders asked me to take on that role, along with strong engineering leaders, especially Kevin Stewart.

The vision had to make it very clear not only to our product teams, but to our executives, stakeholders, investors, analysts, and customers.

We began with our CEO, Shantanu Narayen, and our CFO, Mark Garrett.

We realized that we would be asking these two key leaders to make a very significant bet—potentially a bet-the-company bet—and we knew that if they were not convinced, there was simply no way the rest of the company would go along with such disruptive changes.

So back to the question of what's the carrot? Why go through all this?

The main tool for answering this question was the product vision, and this vision proved to be the key to getting the senior leaders and the many stakeholders on board. We'll discuss the key role of the product vision in more detail below, but the result was that both the CEO and the CFO could understand the motivation and the benefits, and they were able to begin to work through how they could help.

Transforming to the Product Model

While the company faced many challenges, it's also important to recognize that it had several important assets.

Adobe already had a strong engineering and product design culture. It had a strong reputation for technical talent and a track record of providing several of the core components of the technology industry. And perhaps most important, the company had a large and passionate customer base.

That said, the product leadership team and I knew we had to drive significant changes across the company.

Changing How We Build

As our customers' needs were changing in terms of how they wanted to ideate, design, and collaborate, the technologies we used and supported had to change as well, including how we built and deployed our products.

Our waterfall-driven product cycles were simply far too long. It was no longer tenable for our customers to wait 12 to 18 months to be able to take advantage of new capabilities. The market and competition would leave us behind.

The speed of innovation in web standards and mobile devices pushed us toward a continuous development and delivery model.

While Adobe had and still has very strong engineers, early on there was real resistance to moving to the model of continuous deployment.

There is a very obvious trade-off between improving the lives of customers versus improving the lives of the engineers building and supporting products. Historically, if customers discovered a critical problem, they could always just install the previous version. But in a cloud-based solution, if a problem exists, it potentially impacts all customers. There is no easy way for a customer to go to any version other than the one that is live and supported by Adobe. For this reason, the engineers have to constantly keep the service running and working properly.

In addition to committing to this level of service, being able to deploy continuously and independently involves significant architectural changes in how the products are designed, built, tested, and deployed.

To the credit of Adobe's engineers, they stepped up to this challenge and delivered for the company and for their customers. It's just important that readers understand that this was not easy and not without early resistance.

Changing How We Solve Problems

In addition to changing how we built and deployed our products, we also needed to change how we solved problems.

At the time, Adobe had a very engineering- and designer-driven culture.

Engineers or designers typically would come up with product ideas and pitch them to product managers. These product managers would create requirements around the ideas, identifying the problems they solved and the value proposition to the designer customer.

Many of these strong engineers and designers were promoted over time, and engineering leadership was highly valued—in many cases much more so than product management or product marketing.

The good news was that there was already a strong culture of empowered engineers, but the bad news was that there were many layers of people and organization between the engineers and the actual designer customers purchasing our products.

Too often this would result in capabilities that didn't achieve the level of adoption we had hoped for.

We knew that if we wanted the level of continuous innovation that the future would depend on, we needed to empower not just the engineers but true, cross-functional product teams, and then enable them to engage directly with the customers.

So we invested in cross-functional product teams of product managers, product designers, and engineers, working closely with product marketing to engage directly with users to understand their new challenges and to discover solutions that addressed these needs effectively.

While the company had strong engineers and product designers, this change depended on a strong product management role.

To be very clear, there were already many strong product managers in the organization, but historically their role was more focused on building business cases, capturing requirements, and working with product marketing to prepare for the upcoming release. They also had a significant project management responsibility to report progress on the milestones. But we needed the role to change to be primarily about product discovery and the necessary focus on value and viability.

And just as product development needed to move from a project-based waterfall style of work to continuous development and delivery, the product discovery work had to change.

Because the competitive landscape and customer expectations were changing so quickly, we couldn't assume that what was true six months ago was still true today.

Changing How We Decide Which Problems to Solve

A Compelling Product Vision

With engineering, design, product marketing, product management, and others, we started talking about customers and the challenges they had.

We knew many of our customers had an issue with affordability, especially those smaller design shops that couldn't afford to upgrade to the latest version of the tools.

We also knew that most of our customers were very excited about using mobile phones and tablets to ideate and create. In addition, we knew there were many different venues and sites for designers to publish their designs to potentially attract new clients.

Given our market position, didn't Adobe have permission to solve these problems?

Our product vision started taking shape once we realized that instead of focusing internally about what we wanted to accomplish for our business, we should instead start talking about designer customers and how we could improve their lives.

We decided to write an aspirational "day in the life" narrative about a freelance graphic designer we named Marissa.

Marissa had her own design shop in which she employed a few other designers.

One day, the creative director of Nike was browsing creative .adobe.com and came across an interesting piece of artwork that caught his eye. The story goes on to explore elements of the website and allows the Nike creative director to reach out to Marissa to ask her for a proposal for a new advertising campaign he wants to kick off.

Marissa leverages the Adobe tablet applications, cloud syncing, and desktop tools to create a compelling proposal and is also able to source other photographers and designers to help her with the project once she won the bid.

The narrative and the *visiontype* that showcased it became guiding forces behind what we wanted to do for our designer customers.

The belief was that if we did the right thing for our designer customers, then they would stay with us and new customers would join us. We knew we had effective mechanisms to capture the value.

This assumption had to be tested both externally with customers and internally with stakeholders.

We started sharing the vision with key designer customers to explain our view of the future and why we thought it would be beneficial.

Equally and critically important, we started sharing the product vision with key stakeholders. We had a hypothesis that our go-to-market model would have to change, as would how we packaged our offering. We knew that these changes would have significant implications for sales, marketing, and finance.

The visiontype was key in helping us understand not only what the future held for our designer customers, but the current gaps we had in our portfolio, to decide which problems were critical to solve and the priority order to address those problems.

An Insight-Driven Product Strategy

Our product strategy work identified two major problem areas for our customers.

The first problem area was around identity management and the second was cloud-based syncing of data.

Identity was something we knew we needed to address immediately. What might be acceptable on a desktop solution would not be acceptable in a cloud-based solution accessed from a variety of devices.

Multiple identity systems are not only cumbersome for users; they also hindered our ability to collect the usage data that would be so important for both personalization and license management.

We knew we needed to have a common identity system—especially if we were going to have a user access all their content, services, and tools from anywhere.

The second problem area was syncing of files and other data from mobile devices, tablets, cloud applications, and desktop with no loss of data or fidelity.

Product Planning

Remember that the Creative Suite was not just a single solution but rather an integrated suite of more than 20 different products. Moving to a new generation would involve not just moving the component products but also addressing gaps.

We pursued a strategy of migrating some applications, redesigning and rebuilding other applications, and strategically acquiring select components where the technology or the team could help us more quickly achieve our product vision.

The product vision helped in communicating the shared goals of the work and in determining which problems were most important to solve.

Each of the many product teams brought strong opinions on how they wanted to improve their products, and this enthusiasm was good. But we needed the teams to internalize and prioritize *Marissa's needs* and ultimately become obsessed with them.

The vision also required the teams to truly collaborate. Whether it was integrating a common identity system so a customer needed only one Adobe ID, or if it meant that files needed to traverse multiple platforms, a significant amount of coordination and collaboration was required.

Fortunately, we already had this muscle memory because every product cycle we released not only individual point products but Suites that were focused on key customer segments. The product teams referred to this as the *Suite Tax*.

Also, some broad decisions needed to be made from both a technology perspective and a design perspective. Design was a bit easier because we had strong design leadership, and we had a precedent in terms of the belief that customers and users should experience a consistent experience regardless of what device or tool they were using.

The same philosophy held true when it came to cloud-based services and mobile applications. There would be a strong point of view in certain functionality and iconography. A design system would be put in place with fully coded components for where we needed consistency, and then the individual product teams could experiment in areas that weren't required to be consistent.

Changing How We Go to Market

The final aspect of being able to get the Creative Cloud launched was real collaboration with key stakeholders. In particular, we were not only changing our products, but also changing our business model and the product's go-to-market strategy.

Recall that in the past, we typically released and launched a new version of our products every 12 to 18 months. We would have a very long development cycle and a very complex, coordinated go-to-market strategy. We would take months of planning which features would be highlighted, and we would craft messaging and positioning documents, create demos to bring on press tours, and spend substantial time and money on collateral and packaging.

With an SaaS-based subscription model, we would be improving the products and releasing continuously, so we wouldn't have the time to do this.

We would have to change how we informed customers about new features and change our other communication channels. We would need to worry about discoverability of new features because we wouldn't have the marketing machine of the launch to drive awareness of them.

Another challenge was that we communicated with our customers primarily through press and analyst channels. We would go on press tours, where product managers and product marketing managers would showcase and demo the latest releases to key industry pundits to garner favorable reviews. This approach helped customers and users determine why and if they should buy the latest version of the product.

By taking our products directly to the designer customers, we had to ensure that customers could quickly discover and learn new capabilities without the yearly event of installing a new release.

As you can see, this impacts how the sales organization sells to enterprise customers, how the channel organization works with retail partners, how product marketing needs to adapt, and more.

To prepare for all these changes, the product team had to collaborate very closely with the go-to-market teams to manage the transition.

Virtually every part of Adobe was impacted by the transformation, and especially by the changes to the product offering.

If nothing else, I hope I have communicated the scope of the necessary changes. Nothing about this level of change is easy.

The Results

We measured the success of the transformation by the number of customers, the revenue, and the shareholder value.

With the Creative Suite, there were 6 million customers. With the Creative Cloud, by the end of 2021, there were 26 million subscribers.

With the Creative Suite, revenue hit $2 billion a year. With the Creative Cloud, we recorded $11.5 billion in 2021.

During this time, Adobe's market cap grew from $13 billion to $269 billion.

The business outcomes were clear, and this transformation is considered one of the most financially successful in industry history.

PART

X

Overcoming Objections

With any sizable group of people, there will always be objections to any significant change.

It's true that there are always some people who resist change just because they don't like change, and we have techniques for dealing with that.

But the more difficult case is that people often have very legitimate concerns, and they simply don't see how those concerns would be addressed in this new model.

The chapters in this part enumerate the major concerns coming from all across the organization—and also the concerns that come from inside the product organization itself—and then discuss how we address each one.

CHAPTER
36

Objections from Customers

Note: Much of this discussion is informed by chapter 21, "Partnering with Customers."

"Our users believe that it's absolutely essential for you to provide this specific capability. Unless you can commit to that, we can't commit to buying your product."

This is a very common and often reasonable objection from customers.

It's especially important to keep in mind that many of these customers have heard years of promises from companies, and their sales and marketing staff, yet have seen very few deliver on those promises. It is understandable that people would be skeptical after that.

First, it's important to make sure the customer understands the difference between a commercial product company and a custom solutions provider.

Second, realize that it's very normal for customers to express their desire in the form of a solution, and it's the job of product to determine the problem behind the solution they are requesting.

If the customer's situation is truly unique, and represents a market of one, then it's usually most appropriate to direct that company to a custom solutions provider that has a business model designed around this case.

However, if you believe the customer is part of your current or upcoming target market, and you believe this capability is in line with your product strategy, then this is a good opportunity to engage directly with this customer to gain a deeper understanding of what it will take to succeed.

You need to be clear with the customer that you are committed to solving this problem for them, but that you need to solve it in a way that also works for other customers in this market.

Sometimes there is a clear solution that will work for this customer and others, but in other cases, some significant product discovery work will be required to come up with a solution that the customer loves, is technically feasible, and is also viable for their business.

It may be helpful to explain to customers that most of their favorite products were ones that customers did not even realize were possible at the time.

"I need to know when I can count on this feature being delivered."

Per the previous discussion, you need to determine if this feature is part of your product strategy or not. But assuming it is, when a specific date is explicitly required, before you make any promises you might not be able to keep, you need to realize this is what *high-integrity commitments* are for.

"We are making a major bet on your company if we purchase this product, and we need to make sure the direction you are

going is where we want to go as well. We need to see your
product roadmap to know if that's the case."

Especially in the enterprise software world, companies often are
making a very substantial investment and bet on your company.
They need to know not just what you do today, but where you are
going tomorrow.

Most people will ask for a product roadmap, so you may need to
explain the inherent risks of product roadmaps. It can also be helpful
to maintain both internal product roadmaps and external, customer-
sharable product roadmaps.

That said, it is actually the *product vision* that directly addresses this
very real customer need. And it's helpful to the product organization to validate this product vision with feedback from customers.
So, when we have the choice, we prefer to share the product vision
rather than product roadmaps.

"It is so much work to deal with your new releases, I can only
handle them every 3/6/12 months."

"We need thorough documentation and training for every
new release."

These are very legitimate objections, but the root cause of these
objections is actually due to releasing too *infrequently*. You are delivering too much change at once to the users, and you are causing too
much disruption to your customers.

When updating to a new release is a major pain, it's no surprise
people would want to go through this less frequently and not more
frequently.

Sometimes, customers believe that getting a release less frequently
means they will get a higher-quality release. But the evidence is
overwhelmingly the opposite, and you may need to explain to customers why this is the case.

In chapter 4, "Changing How You Build," we discussed continuous delivery techniques to prevent this problem. In the product model, the customer receives a continuous stream of small, reliable, incremental changes. Those changes are designed such that users are not impacted—or to the degree they are, they do not need retraining.

It's also useful to understand that there is a difference between when you release something, and when that new capability is visible to the customer. One continuous delivery technique is to "release dark," which releases the new capability into production, but you control if and when this new capability is visible to customers. This is known as *feature flagging*. *Feature gating* can allow us to control the subset of users that we want exposed to this new capability.

Sometimes, it is helpful to point out to customers that this is already how many of their favorite mission-critical products work, whether it is their browser, their phone, their new car, or even their new washing machine.

If they still believe they need to recertify every week, they can do that, but eventually they will come to trust that you are ensuring they are always running the most stable and effective solution.

Today's customers demand better than the prior era's quarterly or yearly big-bang releases.

"We are very uncomfortable with the idea that you are collecting data on our use of your products. Why should we allow this?"

There are two important things to emphasize here. First, when your products report back usage data to the product team, the data has been anonymized and aggregated so that there is no personally identifiable information. Second, this data is used to make sure the product is working properly for customers and truly helping customers to solve their problem. This is important both for your customers and for you.

Just as a pilot flying a plane depends on instrumentation to show that she is on the right course and that the plane is flying safely, product teams depend on this instrumentation for the same reasons.

"We were working as part of a customer discovery program, and we were very excited about the progress, but then we were told that you decided not to pursue this new product. Is this normal? Did we do something wrong?"

This is not common, but it does occasionally happen, and we hope this possibility was explained to you at the outset. The main reason for this is that the product team was not able to find a single solution that would work across the range of customers they were collaborating with. Sometimes, the needs of a customer truly are unique, and turn out to be better suited to a custom development firm rather than a commercial product company.

CHAPTER
37

Objections from Sales

Note: Much of this discussion is informed by chapter 21, "Partnering with Customers."

"We are the ones sitting down with customers every day. Why shouldn't we be the ones to tell the product team what needs to be delivered?"

There is no question that salespeople operate on the front lines with customers, and as such they are valuable sources of input for the products. In the product model, product managers have strong relationships with salespeople from across the company because they truly depend on one another. Product managers are also encouraged to actively participate in the sales process for key prospects.

More generally, product managers and product teams have direct, ongoing engagement with users and customers. Not in a sales context, but in a product discovery context.

That said, there is a foundational principle which explains why strong product solutions rarely come from the customer or from the salesperson:

The customer and the salesperson do not know what is technically possible.

They are often experts in their own domain, but rarely are they experts in the enabling technologies used to build the solutions they buy.

When they buy your products, they are effectively hiring you to be the experts for them. Strong products come from combining real customer needs with solutions that are just now possible.

"This is a great product vision, but it's years in the making. I need to make my numbers now. How will what's being built help my sales team make their numbers this quarter or next?"

While it is important to share the product vision, it's also critical that you provide the sales force with products they can sell *now*. The best way to do that is to provide them with referenceable customers that are using the latest offering and loving it.

More generally, you may need to explain to the sales leader how each of the product efforts is intended to result in direct business impact.

"I have a very promising prospect, but we need to commit to some new capabilities to close the deal."

In the product model, one of the greatest fears of product teams is what are known as *specials*. These are one-off capabilities that are done for specific customers to satisfy specific requests. It does not take many of these specials before a product is more complex, less nimble, and harder to learn and use for everyone.

The first thing we recommend for handling these specials when moving to the product model is to have any special request (for anything other than currently shipping products) first go through the head of sales. That person can, it is hoped, prevent most of the requests from proceeding.

For the requests that do get past the head of sales, the product manager has a responsibility first to understand if the prospect is in the target market, and then to understand what problem the prospect is trying to solve by requesting this capability.

In many cases, the prospect is in the target market. Once the product manager understands the problem behind the request, she can identify a solution that would meet the needs not only of this prospect, but of other current and future customers as well.

In other cases, the customer is outside the target market, or the solution needed is not within the scope of the product strategy, so the right answer to is decline this opportunity in the name of being able to pursue the larger opportunity.

More generally, this type of special request is very often the result of not yet having referenceable customers, which then forces salespeople into these sorts of conversations with prospects.

"We lost a big deal specifically because we are missing some key capabilities as compared to our competitors. Why would that not be our top priority?"

Every experienced product team knows that customers often say one thing and do another—sometimes consciously but often not.

If a prospective customer chooses a competitor, they will often point to something easy as an excuse, which can lead product teams on a never-ending feature chasing journey.

And sometimes you really are lacking a critical feature. Again, a focus on developing reference customers is generally the best way to know the difference.

"We keep losing in deals against this key competitor. What can we do to better compete?"

In this case, the sales organization is explicitly asking product (and product marketing) for help coming up with an effective response. It's very possible that major product work is underway at the moment to address this, but it's also possible that this work may take awhile, as some major changes may be required.

This is where product managers and product marketing can put their heads together to try to point out opportunities to give the sales team something helpful to say or sell. Are there segments where the current offering is superior to this new competitor? Can we quickly develop some new reference customers for that segment?

"I work very hard to manage my prospects and customers, and I don't want people from product teams risking that."

First, it's important to acknowledge that direct, unencumbered access to users and customers is absolutely essential to creating successful products. It is one of the few truly nonnegotiables in the product model. If necessary, the CEO may need to be called upon to make this point to sales leadership directly.

That said, it is also true that there have been cases of members of product teams saying or doing something inappropriate or unwise while interacting with a customer or prospect. It is the responsibility of the product leaders to ensure their people have been coached on appropriate behavior.

For companies with a strong account control ethos (which is not at all a bad thing), you can ask the sales organization to arrange a short "customer interaction training" for anyone going out to visit customers. This is sometimes referred to as *charm school*.

In addition to learning the roles and responsibilities of the different members of the sales and services organizations—and the expected behavioral norms—it's important to impress upon anyone about to visit a customer and likely to meet users, some of whom may be very unhappy, that any promises they make they will absolutely need to keep.

While this is not an alternative to product teams spending time with customers, it's also worth pointing out that today there are several excellent tools for product teams to get a better understanding of the conversations going on during sales calls, and there is always a great deal to learn both about the product and the product's go-to-market.

"I want to make sure that we don't have random product teams contacting my customers every week without coordinating with each other, and especially without coordinating with me."

This is entirely reasonable, and your product ops/user research team is set up to facilitate and coordinate these customer visits. They ensure that you do not impose on your customers—such as every team wanting to visit the same customers—and also that the appropriate account managers and customer success staff are kept in the loop.

38

Objections from the CEO and Board

Note: Much of this discussion is informed by chapter 26, "Partnering with Executives."

"As CEO, I meet with customers, investors, board members, analysts, and other company executives every day. How would I not be the appropriate person to decide what the product teams should deliver?"

"I am the one responsible for the revenue, so shouldn't I be the one to define the product roadmap for delivering that revenue?"

There is no question that as CEO, your activities put you in a very strong position to drive product. In fact, in most startups, the CEO or one of the other cofounders is usually also the head of product, and this is why.

That said, innovation at scale is dependent on empowered teams, as customers, investors, board members, analysts, and other company executives generally don't know what is just now possible. So, the scalable solution is to push these decisions down to the relevant product team.

But it's also true that some companies never manage to transition to the product model because the CEO does not want to let go.

As long as things continue to go well, that may be for the best. But in most cases, the CEO is forced by the limited hours in the day to decide between running the product or running the company.

"I am trying to run a business. I need to know what will be released, and when it will happen."

Yes, there's no question that as CEO you will need detailed, timely data on the progress of the major product work.

Your product leaders should be able to tell you specifically which product work is expected to contribute to which business outcomes in what time frame. That is what they are doing in their product strategy work.

Also, there will be cases where you will need high-integrity commitments. Just realize that there is an opportunity cost to these high-integrity commitments, so it is to everyone's interest to limit those to the major work that has true inherent deadlines.

Similarly, the product teams depend on you to share the details of the strategic context so they can make good decisions within their teams when you are away on a business trip.

"My board wants to see our product roadmap. I need to provide that to them."

If the board does not understand how your product work will result in the necessary impact to the business, then they will often ask for a product roadmap. But normally, it is not the product roadmap per se they care about. They want to see how the work they are funding in product has the potential to generate the hoped-for results.

In this case, it is up to the product leaders to connect the dots between the product work going on and the hoped-for results.

Before the board meeting, the product leader should be sure to spend enough time with the CEO, CFO, and CRO (chief revenue officer/head of sales) to ensure they all understand how the product work ties directly to the business results, and they have confidence that this work is the right work to generate the necessary results.

It's also important to recognize that if the product organization does not have a track record of delivering these results, then there will be a period where this trust needs to be earned.

"What if I have a product idea? Shouldn't I have a way to share that idea with the appropriate product team?"

In healthy organizations operating in the product model, ideas from anywhere are normal and welcome.

The question is whether that idea is being shared as a *suggestion* for one way to solve a problem, or as an *edict* for what to design and build next. CEOs are often surprised to learn that what they intend as suggestions are often interpreted as edicts. So, it's important to be explicit.

More generally, as Steve Jobs often pointed out, the idea is the easy part. The real work is in seeing if that idea is any good, and then turning that idea into a real product (which he considered *"90 percent of the actual work"*).

"What is a reasonable time frame in which to expect results from our product efforts? How do I know if a product team is struggling or if things are just par for the course?"

To the first part of this question, the product strategy should spell this information out, ideally in written narrative form, so that all key parties understand the reasons for the major product work; how this work is intended to play out over the quarter and over the year; and how the work ties into the longer-term product vision.

Regarding the second part of this question, there are different ways of assessing this. Many companies conduct quarterly business reviews with product teams for this purpose. Others ask their product leaders to dig deep with each product team every quarter and then report on the status and what actions are being taken to help teams that are struggling.

"I want to make sure our organization is in alignment. We need to make sure that product, marketing, sales, service, and operations are all headed in the same direction and coordinated."

This alignment is the reason why many companies using the product model use the OKR tool to ensure alignment on an annual and quarterly basis.

As an example, if we have new products getting released, we want to ensure that marketing, sales, service, and operations are all doing their part to ensure the success of those new products.

"The board says my investment ratios are off for my stage company. They want more investment in sales (direct correlation to revenue) versus product (where revenue is ill-defined). How do I defend increasing resources in product and not in sales?"

If product-market fit truly has been established, then increasing the investment in sales and marketing often is exactly the right move. However, many companies make the mistake of assuming that just because a company has a certain number of employees, it has achieved product-market fit. When the company has not, spending on sales and marketing is typically very inefficient. Several critical

KPIs highlight this point, including how long it takes to sell and the costs to sell, the conversion rate of trial to purchase, and the customer churn rate.

Until and unless product-market fit has been achieved, you need to focus as much of your resources as possible on getting to product-market fit.

CHAPTER

39

Objections from Line of Business

Note: Line-of-business managers are stakeholders but, rather than representing just one functional area of the business, such as finance, HR, marketing, or compliance, line-of-business managers are responsible for a business unit. It may be a small business unit, such as a category manager in an e-commerce company, or a large business unit, in which case the title may be general manager. Many of the issues are the same as discussed in chapter 38, "Objections from the CEO and Board."

"I have P&L responsibility for this product line. So why would I not have control of the engineering resources?"

There is no question that a line-of-business manager is a critical partner in the product model, in very much the same way that a CEO of a startup or scale-up plays a critical role. But just as a startup CEO would depend on her product and technology leaders, the line-of-business manager would want to do the same.

In practice, just as with a startup, many of the most strategic decisions would be done collaboratively, such as product vision, product strategy, and identifying appropriate outcomes for the product teams.

"Leadership expects that I know about anything that can impact my business. I need to know what's coming so I can answer those questions when they come to me. I don't want to look like I don't know what's going on."

Not only is this true, but leadership in the product model depends on informed, customer-focused, data-savvy, engaged leaders.

Chapter 26, "Partnering with Executives," speaks not only to why this is important but also to how we work to keep each other informed.

It's essential that the line-of-business manager share with the product and technology leaders as much of the strategic context as possible.

Similarly, it's essential that the product teams and product leaders share with the line-of-business manager as much information as possible, especially newly emerging insights and learnings from the product teams on the front lines.

Most important, it's essential that product managers and product leaders connect the dots for the stakeholders about how their product work is intended to cause the desired business impact.

"I find it so frustrating that the product team feels they need to repeat all the problem discovery that I've already done before they can get to work on the solution. And furthermore, because of that, too often there's no time left for solution discovery."

This is a common frustration. In hindsight, it should be clear that if you had been able to include at least the product manager and product designer in those customer interactions, you all would have learned the problem space together and the team could get right to solution discovery.

But going forward, sit down with the product team and offer to share anything you can so they can quickly come up to speed on the problem space. Emphasize to them that they do need to understand the problem space in order to discover good solutions to those problems, but also remind them that customers buy *solutions*, so they need to have enough time to come up with a solution that is better than the alternatives on the market.

"We need to move faster. Speed is critical. How can I get these product teams to feel the same sense of urgency I feel?"

Speed truly is essential, but in the product model, we focus on time to money more than time to market. This is also referred to as *outcomes over output.*

If the team builds something your customers don't buy, you will have spent a lot, but gained very little.

The central idea of product discovery is to figure out very quickly and inexpensively a solution that is worth building—one that is valuable, usable, feasible, and viable.

If your teams are not able to test out product ideas in at least one-tenth of the time it would take to build the product, they are doing something wrong.

"The product team is resistant to showing their early prototypes to us as stakeholders, before they think it is of the necessary quality and ready for final review."

As trust gets built between the product team and the relevant stakeholders, this will improve. But anything you can do to encourage the early interactions is good. If the product team thinks you are interested only in final details, such as final colors or fonts, then they will push you off until those are ready. But if they see that you are able to look past the visual limitations of the early prototypes and to understand the direction the product is heading, and if you can provide thoughtful feedback that helps the teams to realize factors and constraints they may not have appreciated before, they will quickly come to value these early feedback sessions.

"I prefer to have on my business team a set of people who report directly to me who can drive product initiatives."

In the prior models, when stakeholders are driving the product and IT is there to implement the features, often the business side ends up hiring essentially its own form of product managers. This is very understandable since the IT organization did not have this role or anything comparable.

But now, in the product model, the last thing you want or need is duplicate organizations—product managers within the product organization and product managers within the business unit.

Normally, when moving to the product model, you would draw from the best product management talent from all sources, but it's critical just to have a single capable and accountable product manager for each product team, and not split that role.

There are cases where product managers do report to line-of-business managers, and then dotted line to a chief product officer. With the right leaders, this technique can work.

But the main argument against this is that you want your product managers to report to product leaders who are skilled at developing strong product managers. Few line-of-business managers have the time or the experience to coach and develop their product managers to the degree that usually is required.

"I run the business, so I should create the product strategy."

At a minimum, the line-of-business leader would collaborate closely with product leaders on product strategy.

That said, usually the intention of the product strategy is to look holistically *across the business units* and work to maximize the value of the business as a whole.

So, product leaders would work with each of the business unit leaders to ensure they are pursuing the best opportunities across the company, not just within each business unit.

As discussed earlier, this requires trust and transparency.

CHAPTER
40

Objections from Customer Success

"Every day we deal with struggling customers. We are literally on the front lines. Yet the product teams seem not to care about what we think we need."

First and foremost, if a product team truly doesn't care about your needs, then that is a serious problem, and you should immediately bring this up with product leadership.

That said, realize that product teams usually are getting pressured from every side to improve a product—directly from customers, from sales, from marketing, from operations, and from senior leadership, to name a few. So, it is very possible that the product team does care, but it can be difficult for them to hear the signal through all the noise.

One technique that we encourage in this situation is for the customer success team to keep a single, live top-10 list of the ten most serious issues that are causing pain for your customers.

For many organizations, this list can be derived from the tools used to track issues, but it is fine if this is a subjective list based on your judgment.

This top-10 list does a great service to the product teams because they know that if an item is on this list, it is a serious issue, and they don't have to bother weighing it against everything else.

It's important to realize that there will *always* be a top-10 list, but product teams should always be working to remove those items.

It's also important to realize that sometimes product teams will address the problem with a solution different from how the customer or your customer success staff may have imagined. This is because, in many cases, the "obvious fix" has undesirable consequences, and the team must seek a different approach. This is normal, and all good, so long as the problem gets solved.

"The tools we have to take care of our customers are pretty terrible. This not only makes our job harder, but it impacts our customers. We say we're supposed to care about our customers, but how do we address this?"

Today, commercial tools available for customer success teams are better than ever before, so there's little excuse not to be using them. That said, there are areas that are not supported by commercial tools, and that's likely what we are talking about here.

If the actual customer experience is impacted, then the tools you are using would be considered product (even if the end customer can't see the tool).

It is true that many companies do struggle to staff these non-customer-facing products with the level of product team they would like, so you may have to be vocal about the importance of improving these tools.

"We are talking to customers all day, and if we don't know when and how things are going to be changing with the product, how are we supposed to know how to help customers?"

One of the consequences of very frequent, small, reliable releases is that often things are changing, and it can be hard to keep track of when certain capabilities have been deployed or made visible to customers.

This has always been a challenge for companies, but with continuous delivery, the problem is exacerbated.

To address the problem, product teams communicate with product marketing, and then you depend on product marketing to communicate with customers, sales, and customer success whenever anything becomes visible to customers or requires a change of behavior by the customer. The book *LOVED* describes several techniques to help with this.

Occasionally things will slip through the cracks. It's important to understand why, so that the problem can be prevented in the future. But the major benefit of this approach is to significantly reduce, if not eliminate, disruptive changes for your users and customers.

"Our product organization frankly takes too long to get products out, so our customer success team has staffed our own people to create solutions for customers."

While we applaud the initiative of a customer success team that takes things into its own hands on behalf of customers, this solution simply never ends well. Very soon, customers are running an unsupported blend of some products and some custom solutions—held together with lots of duct tape. Something new gets released and the nonproducts stop working.

If your customers are forced to run a mix of supported and unsupported products—cobbled-together half-solutions—both your customers and your company lose.

The right fix is to improve the product organization and the products produced, which is one of the reasons why companies move to the product model.

41

Objections from Marketing

Note: For the purposes of this chapter, we are combining all the different types of marketing—including product marketing, field marketing, corporate marketing, and brand marketing—into a single chapter.

Note also that much of this discussion is informed by chapter 23, "Partnering with Product Marketing."

"We interact with sales every day, we continuously monitor the competitive landscape, we conduct focus groups with current and prospective customers, and we have relationships with all the major industry analysts. Who better to define the products we need to succeed?"

Indeed, this was how products were created in business software companies two decades ago. The reason companies moved on to the product model was because innovation was so rare in the prior model. That's because marketing, sales, customers, and even industry analysts don't know what's possible.

Steve Jobs famously captured this reality by holding up his iPhone and stating *"You can conduct 100 focus groups, you'll never get an iPhone."* Today, most marketing organizations understand this.

That said, marketing does have very valuable data and insights, and product marketing managers work hard to get any information they consider potentially valuable into the heads of the relevant product managers and product leaders.

"How can we best help the product teams to create successful products?"

Product teams depend on many others across the organization, including sales, marketing, operations, and customer success, to name the four most obvious partners.

Product marketing works closely with product teams, especially on the market-fit aspects of achieving product-market fit. But more generally, they work on sales enablement, messaging, positioning, evangelism, growth, and more. In certain cases, a product marketing manager is even embedded into product teams.

But the main thing to keep in mind is if you believe the company has to fix the product to fix the problem, then you need to ensure you get whatever information is available into the heads of the relevant product managers who are responsible for changing the product.

"Marketing wants to promote a future state of the product in advance of it being done. When is this okay?"

There will be times when it makes sense to explain to the market (and board members, investors, and prospective employees) where you are heading *before* you are shipping those products.

You do have to be careful because it's very easy to cannibalize your own sales, or create overwhelming pressure on your product teams before your new products are ready, or even truly discovered.

You also have to be careful not to box yourselves in with claims and directions that you later realize are not what you need to do.

The key is to not do this before you have the evidence that you will be correct. Whatever you do needs to be done in very close coordination among the head of product, the head of product marketing, and the chief marketing officer.

CHAPTER

42

Objections from Finance

Note: Much of this discussion is informed by chapter 24, "Partnering with Finance."

"We need more flexibility in our spend levels. We want to be able to increase and decrease our technology investment depending on business results. Managing projects with outsourced vendors helps us do that."

"We need to keep our costs down, and the loaded cost of an outsourced engineer is significantly less than the loaded cost of our own employees."

If you look at the loaded cost on a *per-engineer* basis, then you will often find that outsourcing appears to be less expensive. But if you look at the loaded cost on a *per–product team* basis, the results reverse.

A smaller team of *employees* usually outperforms a significantly larger team of outsourced staff. This is not a claim about the relative merits of the people involved. It is a consequence of the role and working relationship.

And if you look not at *projects* but at achieving *business outcomes*, outsourcing is dramatically more expensive. In fact, most outsourcing firms refuse to even sign up for outcomes.

Earlier we talked about there being two sources of value once you build something. There is the thing that you built, and there is the learning that accrues from building that thing. When the people are constantly changing and not contributing at more than an order-taker level, you do not get the value from learning you depend on.

There are some occasional situations where temporary staff can make sense, such as for a specific integration or for a burst of work like for test automation, but, in general, if technology needs to be a core competency, this area needs to be insourced.

More generally, it's important to understand that team stability plays a significant factor in both productivity and morale. It can be very disruptive and expensive for a member of a team to move to another team and need to learn the team's technology, problem and solution space, customer issues, and more. Normally it is better for you to expand the charter of each product team so that you can bring work to the relevant teams rather than move employees to the new work.

"We need to identify any potential projects for the year and understand how much each potential project will cost as part of annual planning. That is how we can decide which ones to fund and move forward with."

"We need to be responsible with our money. We need to know what we'll receive for each project we fund. Otherwise, how can we possibly know what is a good investment and what is not?"

This argument is definitely logical, but you probably have already figured out that the business cases you've historically been provided have been far from reliable in predicting returns. In fact, very few finance organizations even try to hold their people accountable to what they promised.

This is why one of the most important lessons that product teams learn is to know what you can't know.

With technology efforts, it's very difficult to know what a given effort will cost. It's even harder to know what revenue actually will be generated, as that depends entirely on how good the solution is and whether customers choose to use or buy it.

You may notice that this situation is the same for professional investors of technology efforts—venture capitalists. They handle these unknowns by investing a small amount (seed funding) for the company to do their product discovery work. For those that show traction (demonstrable progress), then venture capitalists invest more substantially.

More generally, in the product model, instead of funding *projects*, you fund *product teams*, and you hold them accountable to product outcomes (business results).

"How do we hold product teams accountable for the results they commit to?"

In general, you assign each product team one or more key problems to solve, and then you identify the key measures of success. You also tell teams how conservative or aggressive you would like them to be in pursuing solutions. This is part of the product strategy from your product leaders.

Some teams might be asked for numbers that are very conservative and they are nearly certain they can deliver on, and others might be asked to be much more ambitious in their risk taking, knowing that it's unlikely they will succeed in this quarter. The accountability is a function of the certainty you ask of the teams.

More generally, realize that each product team is contributing to a larger product strategy, and you hold product leaders accountable for the results of that product strategy.

The product leaders are essentially placing a series of bets, and they succeed or fail at the end of each quarter or year based on the overall business results they are targeting.

In any case, if and when a team clearly fails to deliver on one of its commitments, at a minimum, you conduct a postmortem—looking at the causes and potential remedies to prevent this issue from recurring in the future.

"We need to be able to move resources between efforts as business needs dictate in order to make the best use of funds."

This comment also sounds logical, but, in practice, there is so much time needed to come up to speed on a new set of technologies, new types of customers, and new working relationships that shuffling resources turns out to be a very inefficient use of funds, and it dramatically reduces the team's ability to achieve outcomes as well as your chances of innovation.

Instead, we prefer to staff product teams (as opposed to project teams) with a broader charter. Then those teams do all the work for all the projects and features related to their area.

This approach keeps product teams stable, and leverages their growing expertise and efficiencies.

"We need to see consistent positive progress every quarter to know which efforts we should continue to fund. Our job is to monitor that progress and impose that discipline."

Again, while this approach sounds logical, the curve for technology-powered products has a different shape. Experienced product leaders manage the product teams as a series of bets placed on specific technologies, specific people, specific skill sets, and specific data, and adjust each quarter based on the prior quarter's progress.

"We are trying to understand how to treat product discovery from an accounting perspective. Is it research? Or is it just the definition and design stage of product development?"

Discovery is *not* true research (R&D).

Product discovery is an integral part of product development, analogous to the "definition and design" of earlier models. Product discovery is fleshing out the work that will be built and delivered in the coming *weeks*.

When true research is needed, it is done outside of product discovery.

43

Objections from HR/People Ops

"These new job competencies are very disruptive to our company. We have many people in roles that no longer apply, and others where the scope of responsibilities has changed significantly. Changing job classifications is not minor, and there are considerable costs involved. We would rather leverage our current talent and stay with our existing job classifications."

"We have many people in roles that are close to technology, so rather than hiring product managers, can't we just change their titles to product management and send them to some training?"

We understand that this change can be disruptive and potentially expensive. We also understand how the differences in the jobs might appear minor, given the legacy titles. But there is a reason why we say this is very often a major reason that transformations fail. We have never seen a company succeed when it insists on leveraging the current people and job classifications.

It is also important to realize that not all your people will even want the new product manager role. It is a very demanding role, especially compared to the product owner or business analyst role.

That said, if you have managers who are willing and able to do the necessary coaching (admittedly a big *if*), and if you have people who are eager to learn, then our first choice is to try to coach and develop these people. As a rule of thumb, a capable manager should be able to coach a reasonable candidate for a product manager role to competence in three months. If, after three months of sincere effort, the person still is not able to do the job that is required, then you need to find that person a different role and find someone capable to replace them.

"We already have annual performance reviews, and now we're told we need to introduce ongoing assessments and coaching plans. Isn't this redundant and also very expensive?"

Ongoing assessments and coaching plans are there to help your people learn the skills necessary to work in the product model. This is the primary responsibility of your first-level managers, and this coaching normally represents the manager's major responsibility. So yes, this is quite expensive, *but it is not nearly as expensive as having all this staff who don't know how to succeed in their jobs.*

It is worth noting that companies that have a strong investment in coaching have higher employee retention rates, and can hire more staff directly from universities at lower cost.

"We already have an onboarding program for new employees. Why do we need another onboarding program for product people?"

Normally, these two types of onboarding programs cover very different topics. Everyone can benefit from a company onboarding program, but product people also have very role-specific training and education.

"We have annual budgets for raises and promotions. Yet now we are talking about incentivizing managers to work to get their people promoted. How will this be paid for?"

In the product model, you get extraordinary value from employees (especially product managers, product designers, and engineers) who can take on additional responsibilities, such as especially tough problems. Getting a larger number of people truly qualified for promotion up the job curve would be a very good problem to have.

Managers tell their people that it is their job to get them ready for a promotion, but the company still will need to have roles and increased salaries available before promotions can be granted.

"We want every employee to have individual OKRs."

For most roles in a company, having individual OKRs is fine. But members of product teams—especially product managers, product designers, and engineers—need to be incentivized to work *together* as a product team, collaboratively, on the very same problems with the very same measures of success. One of the main ways you do that is to assign OKRs to the *product team* rather than to the individual members.

If you still want each person to have personal goals, this is much better incorporated into the coaching plans.

44

Objections from the CIO

"I come from the school of thought where technology exists to serve the business. In fact, I pride myself on my ability to deliver on the needs of the business. I'm not sure where this move leaves me."

Moving to the product model is essentially about changing from viewing technology as a cost center to viewing technology as a profit center.

We have worked with many CIOs who have expanded their skills to serve in this broader role (sometimes titled as a combined CTO/CIO). But these are the people who were eager to make this change—in fact, in several cases, they were ones who drove the move to the product model.

It's important to point out that the product model is not just concerned with the technology that your customers interact with directly. There is usually a substantial amount of engineering work on technology behind the scenes—either in platform services or customer-enabling tools or internal tools.

We realize there are some CIOs who do not wish to change. In this case, they would stay with the "true IT" vendor management and systems that run the business, and someone else is brought in to lead the broader product engineering.

There are of course several similarities between the old model and the new model when it comes to technology, but there are also some important differences:

A product engineering organization is primarily there to *build*, while a true IT organization is usually more about integrating vendor-supplied systems (*buy*).

There can also be some significant differences in scalability and performance requirements for the two types of systems. (IT supports hundreds or maybe thousands of users, while product engineering often supports systems one, two, or even three orders of magnitude larger).

These differences usually drive differences in engineering job descriptions and pay scales.

Another very important difference is that in true IT, outsourcing is encouraged as true IT is not the company's core competency. But in product engineering, outsourcing your engineering is generally fatal to any chance of innovation.

"I'm currently a CIO but I want to lead this change to the product model, and I'd like to expand my role to CTO. Where can I learn more?"

Some excellent product and engineering leadership coaches can help you with this transition.

There are also excellent CTO bootcamps designed to share best practices and the latest techniques.

CHAPTER

45

Objections from the PMO

Note: PMO *here refers to the program management office, which is essentially a project management function with senior-level visibility.*

"My understanding is that many of the marquee product model companies have a project or program management function. What role do they play in the product model?"

Not only do many marquee companies have the project or program managment function, but the function plays an essential role in the companies. When building large, complex products, especially eco-system products, there are many moving pieces and there are a very large number of dependencies and impediments. The project management role is key to managing these dependencies. The issue is not with the role; it is with the culture behind how the company provides that role.

The problem comes when the PMO, under the guise of reporting on status, exists to instill and govern a command-and-control culture.

When the PMO is there to provide servant-based *delivery manage-ment*, it is providing a major dependency tracking and impediment removal service to the teams, but in a way that is empowering rather than disempowering.

"I come from the school of thought where everything is about predictability. We need to focus on delivering what we say, when we say. We let the business leaders worry about what gets built, and we focus on being a reliable and trustworthy machine for delivering what is requested."

This is a good summary of the old model. But in the product model, the focus is no longer on predictability but rather innovation. Pre-dictability is important only if it delivers the value the company depends on. So, the product model is based on a very different set of goals and priorities, often characterized as time to money rather than time to market.

This is why the PMO often has a difficult time with the change to the product model.

In many companies, the people in the PMO can make this cultural change and serve as effective delivery managers, but in others they cannot, and some decide to leave.

This is one of the more difficult situations in transformation, because in most cases, the people are there to do a job, and there are just dif-ferences in how that job is done. But in the case of a PMO, this is more of a cultural and process view that's largely incompatible with the product model.

"What are we supposed to do if we are no longer playing the program management role?"

Normally, you change the PMO to a delivery management organi-zation. The focus is now on servant-based project management and impediment removal. With the right mindset and skills, delivery managers are very valuable to the organization.

While old-style PMO is perceived as the enforcer of "the business," delivery managers serve product teams, and product teams serve customers, in ways that work for the business.

Just be very careful that you have people who truly understand and embrace this difference and are not merely biding time until there's a return to the older model.

CHAPTER

46

Objections from Inside Product

Note: For the purposes of this chapter, these objections represent the broader product and technology organization (product management, product design, engineering, product ops, and product leadership).

"How can we be responsible for outcomes if we don't control all the people necessary for delivering on that outcome—sales, marketing, services, and so on?"

This is a very common objection, and it's also very understandable. One way of addressing it is to focus the key results on product outcomes that are largely in your control. But you do need to be careful here because it's important that the product team take some real responsibility for the success of the product. The original motivation for OKRs was to encourage product teams to get out of the office and figure out what needs to change. Is there a problem with

312

the marketing or with the sales? Maybe the sales tools? Or is the problem with the product not doing what it needs to do in the customer's environment?

Think of it this way: Who in the entire company has more of an ability to impact the products than the product teams? Think of the salesperson who depends on selling the product for their livelihood, yet has little to no control of the actual product.

"It is very expensive for us to do a release of the necessary quality. How can we possibly afford to do these releases much more frequently? Throwing more people at the problem doesn't necessarily speed the process up—it can actually slow it down further."

True enough, more people are rarely the answer for smaller, more frequent releases. This is definitely a case of working smarter, not harder. There is a saying in the engineering community: "If it hurts, do it more often." If you're doing monthly releases, it is going to hurt. But if you push yourselves to move to weekly releases, or even to daily releases, you will be forced to invest in your test and release automation, and soon you will remove that pain.

The book *Accelerate* goes deep into the theory of why releasing frequently is both faster *and* results in higher quality.[1]

"We are in a regulated industry. We're not allowed to test ideas on our customers, and we're not allowed to release this frequently."

"We are in a regulated industry. We need to follow formal processes in order to be compliant and responsible."

---

[1]*Accelerate: The Science of Lean Software and DevOps: Building and Scaling High Performing Technology Organizations* by Nicole Forsgren, Jez Humble, and Gene Kim (IT Revolution Press, 2018).

It is remarkable how many regulated companies think they operate under certain constraints that the actual regulations don't describe. There are different rules, and you need to read the relevant regulations yourself, then talk to your legal and compliance people to understand exactly what the constraints are, not just what's always been done. But we routinely get customers to sign documents indicating their understanding and agreement to running an experimental version. And, in fact, if you genuinely care about your customers and keeping the technology they depend on reliable, then you *need* to release small, frequent releases.

But, in general, it's common that product teams can do much more than they think they can.

"We don't have time to test product ideas. We just need to code and ship. We are fine letting the stakeholders decide what they want us to build."

If this approach worked, you would not need the product model. The irony is that this objection usually comes from engineers. Yet, in strong product companies, engineers know that they are the primary source of innovation, and they care just as much about *what* they build as *how* they build it.

Your engineering leaders will need to actively coach the tech leads to make sure that engineers are in place who understand this.

"Some of our engineers say they just want to be told what to build. Is there room for those people on a product team?"

The short answer is yes. You do need to ensure that at least the tech lead cares as much about *what* they build as *how* they build. But if some of the engineers on the team just want to build, that's fine.

The longer answer is that companies operating in the product model understand that engineers are the key to innovation, and so they focus their hiring on the types of engineers who don't just want to be told what to build.

"If we test our product ideas with real users, then our competitors will find out what we're building."

First, companies that do product discovery only end up building a fraction of the ideas—the ones that work out well. So, anyone watching wouldn't know the results and what you decided to do. Second, if you are still nervous, you can have users sign an NDA.

"What if we show an idea to a customer in product discovery and they get all excited, but then we decide not to build that?"

This is a possibility, and you should explain to people you test with that you're just learning about what the best solution would be. But, more generally, you do this testing so that you can deliver valuable solutions to your customers.

"When we do product discovery, we sometimes discover opportunities we think are a better use of our time than the problems we've been asked to solve. Why can't we pivot to pursue the larger opportunity? Isn't that what empowered teams are supposed to do?"

First, if you discover an important new opportunity, you should absolutely raise it to your leadership. This opportunity may in fact be what the product strategy should focus on next.

However, the current product strategy likely depends on you solving the problem you were asked to solve. So no, you can't just change what you're working on. Empowerment does not refer to working on whatever you like. It refers to being able to come up with the best solution to the problem you've been asked to solve.

"We are trying our best to please all our various stakeholders, but it literally seems impossible to give them all what they want, as well as deliver on the outcomes we're expected to achieve. It feels like we're being set up to fail."

It is true that sometimes solving for all parties is especially difficult. Some of the innovation stories in this book have highlighted such cases, and you can see how the team worked through the conflicts. But normally, product teams have one or at most two problems they're trying to solve, plus their keep-the-lights-on work. If you still feel you're being set up to fail, this is important to discuss with your manager. Remember that in team objectives, the product team says what they think they can accomplish. So, you are very much in control of this.

"So much of what we want to do depends on other product teams, and we can't control what their priorities are. Even small things take too much coordination work. This situation certainly doesn't feel empowered."

This is a common consequence of a poorly designed team topology, and also of a large amount of technical debt.

Nevertheless, since it is very common, it's important to discuss what you can do about the situation now, while simultaneously addressing your tech debt and considering more substantial changes to your team topology.

First, consider the possibility of increasing your team size, as it's likely that a smaller number of somewhat larger teams would leave everyone feeling both more empowered and more autonomous.

Second, this is where delivery managers can be a real help. They can help track—and, when possible, resolve—the various dependencies and impediments.

Third, you may want to consider expanding the use of platform teams. Platform teams do introduce explicit dependencies on them, but these dependencies are easier to manage and usually end up substantially reducing the total number of dependencies. This is often a step toward the necessary tech debt replatforming as well.

"We have a hard time understanding what decisions we are allowed to make ourselves, which we are supposed to propose

as a recommendation for approval, and which are simply ones that will come down from above."

This is a very normal question to ask your manager during your weekly 1:1. Failing to do weekly 1:1s is just another sign of much more serious issues.

In general, when you consider decisions, you'll talk about risk and consequence. What is the risk, and what is the consequence of making a mistake? In many cases, the consequences are minimal and reversible, and as such are usually handled by the team.

A simple and useful way to frame product decisions is Amazon's one-way (nonreversible) and two-way (easily reversible) door decision metaphor.

"Sometimes we don't agree as a product team what the best decision is. How are we supposed to resolve these disagreements?"

First, it's a normal and even healthy sign that the product team has different opinions and feels safe enough to share them.

If the decision pertains to a specific area of expertise, you'll normally defer to the person with that expertise.

If that doesn't resolve the dispute, the usual answer here is to run a quick test. Learn what the best answer is together.

If you're unsure, this is a perfect question for a weekly 1:1 with your manager so that you can discuss your specific situation.

"We need a central role to control what all our commitments are and what all the various dependencies are and expected dates. If we don't have a PMO, how will this be done?"

It is very common for the head of engineering to review and personally approve all high-integrity commitments because ultimately their reputation is on the line.

As to tracking dependencies and related dates, you should have delivery managers who help you with this and who help remove impediments that get in the way of delivering.

"A specific product team keeps complaining that it takes most or all their time to just do the keep-the-lights-on (KTLO) work."

This problem is unfortunately common and very real. There are a few factors to consider: What is the actual level of KTLO work? Is this an ongoing issue or just temporary? Is this a platform team (there to enable other product teams) or an experience team (either customer-facing or customer-enabling)?

What percentage of the team's time is going to this keep-the-lights-on work (up to 30 percent is normal for an experience team and up to 50 percent is normal for a platform team).

If the situation is ongoing, and if the average is above the guidelines just listed, the normal fix is to add one or more engineers to the team so that the overall percentage of keep-the-lights-on work is back in proportion.

If adding engineers is not an option, then since KTLO work is, by definition, not optional, you would need to reduce the non–KTLO work for this team. Doing this will negatively impact morale as well as this team's business impact, but it may be unavoidable until additional help is found.

"Some engineers believe it would be faster for them just to build something in order to test it."

Sometimes this is true, and it's a win when it is. But more often than not, engineers who believe this have not been exposed to or trained on the modern techniques to very quickly create testable prototypes.

But assuming the team is skilled, then if the engineers can build something in production faster than can be prototyped in discovery, and so long as those engineers are okay with throwing the code away if the product idea doesn't work, then this is fine.

"The engineers are complaining that the product manager and designer are not including them in their discovery work other than to tell them what to build in the end."

This is one of the most common problems, and the result is not simply lower morale on the part of the engineers. An even larger issue is the lack of innovation that happens in this model. We encourage engineers to first speak about this directly with the product manager and product designer. If that doesn't help, this is one of those occasions when an escalation to a product leader is important.

"The team is complaining that they're being asked to provide dates for everything."

This situation is usually correctable with some discipline. First, make sure that everyone knows there is a procedure that is always followed when dates are needed. This procedure is called a *high-integrity commitment*, and it requires time from the engineering tech lead and potentially additional engineers.

Second, make sure everyone understands that these high-integrity commitments take some time to produce—there is an opportunity cost to coming up with this date.

Third, work to educate the organization as to when a date is truly necessary versus when it is not. Explain that high-integrity commitments are meant to be the exception and not the rule.

"The team is struggling with so much technical debt that even small items become major, taking a big toll on morale as well as results."

This very serious situation will require senior product and technology leaders to work with the executive team to get a recovery plan in place. Specialty firms can help with your plan for digging yourselves out of this tech debt hole, but doing so is still painful, and many companies don't survive this situation.

There is too much involved to discuss here, but normally it will take one to two years to get back on track, and that's if things are managed well.

"The team thinks there's no time actually to test the risks before building."

In the product model, we often talk not so much about time to market but instead time to money. In other words, what matters most is achieving the necessary results. If you don't test the risks before building, you will very likely spend the next several months building something that will not achieve the necessary results. Then you'll have to go through the same cycle again, usually multiple times.

In contrast, testing those issues with prototypes instead of products—in hours and days, instead of weeks and months—normally will allow you to achieve time to money much faster than otherwise.

More generally, it is not unusual, during even the first hours of product discovery, for the product team to realize that a solution is not good and is not worth building, which saves substantial wasted effort. It's worth noting that without this product discovery, the team would have to spend the time and money to build this solution only to find that it was not a good use of time. For this reason, most experienced product teams would argue that you don't have time *not* to consider the risks in discovery.

"The discovery work and the delivery work are out of balance. Either the discovery work can't keep up or the delivery work can't keep up."

It's not unusual for product teams to get out of balance temporarily—perhaps a week or two at a time—but if the problem is sustained, then it's typically a sign of either too few or too many engineers.

If the engineers can't keep up with the discovery work (there is too much work on the product backlog), then it's typically a sign of too few engineers. If the discovery work can't keep up with the engineers, it's usually a sign of too many engineers.

"The team is struggling with remote employees. Work is taking an unusually long time; team members don't feel included; psychological safety is dropping."

Many teams struggle with remote workers, and two techniques are effective for improving the situation:

First, arrange to spend some time working in person at least once every quarter. The location you select is less important than the duration and frequency of the in-person visit.

Second, increase the 1:1 coaching for the people who are struggling. Consider twice-a-week, 30-minute 1:1s.

"The engineering leader keeps moving people from team to team based on weekly or monthly needs and is not appreciating the importance of durable teams."

While it is possible that the product or engineering leader simply doesn't understand the importance of team stability or durability, more often this is a sign of a problem in the team topology.

If a product team is scoped too narrowly, then teams will struggle with sizing. Instead, it is better to have larger teams with broader charters, and let the product teams themselves worry about how they pass around the various work to one another rather than move people between teams.

"We have all these other people who were not covered in your product model competencies. They include product owners and business analysts. How do they fit in the mix?"

The short answer is that they don't. The product model does not have these other people on the product team.

The product owner is a role, not a job, and that role is meant to be covered by the product manager.

In the product model, the business analyst responsibilities move partially to product managers and partially to product designers.

47

Innovation Story: Kaiser Permanente

Marty's Note: Many people believe that it's effectively impossible to innovate in regulated industries, or in large, complex organizations, or in companies with many different stakeholders. I love this example because it proves all these people wrong. While by no means easy, with the right leadership and motivation, innovation absolutely is possible, as you'll see here.

Company Background

Kaiser Permanente is one of the nation's largest nonprofit healthcare and health coverage organizations, serving more than 12 million patients.

In 2019, Kaiser Permanente's digital organization embarked on a large-scale transformation to the product operating model.

The organization built out a digital product and technology group with an initial focus on reimagining the patient health journey and streamlining the patient's digital health experience.

Product teams organized around the patient journey, and those teams helped automate manual processes and streamline digital experiences, resulting in improvements in patient satisfaction, better adherence to medication and care plans, and improvements in operational efficiency.

Then, within a year of launching the transformation effort, the COVID-19 pandemic began.

Almost overnight, Kaiser Permanente's care delivery approach had to change due to the public health emergency and the significant limitations impacting in-person care.

The Problem to Solve

The company had certain telehealth options in place for several years. However, the existing telehealth solutions had some fundamental constraints.

The existing telehealth offering was available only during normal business hours, requiring patients to visit in-person urgent care centers and emergency departments for after-hours care.

Patients could schedule a telehealth appointment only in their home markets, preventing them from receiving virtual care when away from home.

A combination of technical, regulatory, clinical, and operational constraints prevented telehealth from being offered in a consistent manner across the company's many markets.

The potential for care gaps to increase during the pandemic was very real.

Kaiser Permanente urgently needed a way to expand its care delivery approaches so patients could receive the care they needed.

The company needed a technology-powered solution that would enable patients to get access to the right care, at the right time, no matter the market they lived in, or their ability to get to a medical office, or the time of day.

This effort became known as *Get Care Now*, providing a single, nationwide, 24/7, on-demand virtual care solution for patients.

Discovering the Solution

The product organization understood that there were very substantial product risks involved.

The priority and motivation were to meet patients' needs during the increased health risks brought on by the pandemic.

From the patient perspective, there would be important usability risks, as patients bring a wide range of comfort with technology, and they may also be experiencing substantial stress during the times they need care.

Patients needed both to understand how to use the service and to decide to use the service over what they were already familiar and comfortable with: visiting a clinic or emergency room.

From the perspective of the care provider (physicians and medical staff), these clinicians needed to learn and understand this new avenue of care and be sure they could deliver the necessary patient experience before, during, and after the telehealth visit. Beyond the actual experience clinicians provide to the patients, very specific medical, regulatory compliance, and operational needs exist.

As the online experience would be new for patients and clinicians, the product teams needed to learn the expectations and behaviors of both groups as they interacted virtually. How long would patients be willing to wait for an appointment? What conditions would be best suited to a virtual session, and which would require an in-person examination?

The product teams collaborated closely with clinicians and business operations to explore potential solutions and approaches that would meet the needs of patients and care providers, as well as operational and compliance needs.

To keep things simple and intuitive for patients and to prevent long wait times, the product teams would need to solve some very complex challenges on the operational side, such as making sure a physician licensed to provide telehealth in that specific regional market was available. In terms of infrastructure, the company needed to ensure clinicians were equipped with the technology necessary to securely interact online with patients from either a medical office or their own homes.

Product teams worked side by side with clinicians and operations, identifying critical operational constraints related to which clinicians were staffed as well as how they documented the care interactions and enabled handoffs between clinicians. In addition, clinical protocols around acuity (severity of illness), triage, assessments, and documentation were identified and incorporated into the patient and clinician experiences.

The key to the critical solution discovery was the product teams partnering closely with the clinical and operations teams in each of the company's regional markets. The teams employed continuous ideation and rapid testing to converge on solutions that addressed the needs of each of the critical parties.

As the many operational issues were addressed one after the other, the teams focused on the remaining technical feasibility risks. Critical challenges existed in integrating underlying scheduling and medical health records systems as well as clinician tools.

Further, with more than 12 million patients, the solution needed to be truly scalable.

Key data around clinician capacity, appointment schedules, and patient health information needed to be extracted and aggregated to support *Get Care Now*.

Once the product teams believed they had discovered and validated an end-to-end solution that addressed the many product risks, they built and quality-tested the product, began a gradual rollout of the solution across all their existing markets, and later expanded nationwide.

Remarkably, *Get Care Now* was created in only four months from inception to launching the new experience across the company's markets. With actual usage data being collected from patients and clinicians after deployment of the experience, the next iteration was implemented and deployed three months later.

The Results

Since the full launch of *Get Care Now*, Kaiser Permanente's patients have had 24/7 access to virtual care from their care providers online in every state nationwide.

Most important, member satisfaction was 9.6 out of 10, with a remarkable 88 percent net promoter score. More than 36 percent of members using the service are seen after hours—between 5 p.m. and 8 a.m.—saving many of these patients from needing emergency room visits and reducing the strain on that resource.

More generally, too many people believe that in regulated industries like healthcare, it is impossible to do this level of technology-powered innovation in any amount of time, let alone in just a few short months.

Kaiser Permanente was able to demonstrate this ability to themselves, to their senior leadership, and to the broader industry, providing real value to their patients and to their many dedicated care providers.

PART

XI

Conclusion

The purpose of this part is to connect the dots between all the critical points discussed thus far, including the common themes from the successful transformation case studies.

CHAPTER

48

Keys to Successful Transformation

In this chapter, we want to share what we have come to believe is essential in order for a company to effectively and successfully transform.

We have discussed each of these key concepts in this book. Everything we have described is meant to increase your chances of success, but it's also true that not everything is equally critical.

Here we put the focus on the ten things we consider to be most critical for increasing your chances of success.

1. The Role of the CEO

While theoretically not impossible, it is extremely difficult to transform successfully without the active support of the CEO. The other items on this list will make obvious why this is the case.

And to be clear, when a CEO decides just to designate some leader as being responsible for "digital transformation," that is not what we are talking about. This common mistake makes it all too easy for the rest of the company to continue business as usual.

Understand that while product management, product design, and engineering may be at the center of this transformation, the impact necessarily goes far beyond the product organization.

It's not unusual to find that you also need fundamental changes to finance, human resources, sales, marketing, customer success, business operations, and more, and occasionally the CEO will need to get involved.

More generally, the CEO really needs to be the chief evangelist of the product model.

2. The Role of Technology

At its most fundamental, a successful transformation changes the role of technology from a necessary cost to the core enabler of the business. This mindset impacts nearly everything—from how technology is funded to how teams are staffed, and whether technology is considered a core competency or something that can be outsourced.

3. Strong Product Leaders

Assuming the necessary support from senior leadership, and that's a very big assumption, then everything depends on strong product leaders. Specifically, these are the people who lead product management, product design, and engineering.

The importance of this can't be stressed enough. *These are the people who are responsible and accountable for everything that follows.*

You'll need to be sure either that you have experienced leaders who know how to work effectively in the product model or, at the very least, that they have product coaches to help them through the upcoming transformation work.

4. True Product Managers

Empowered product teams depend on competent product managers. For most companies that are trying to transform, even if they currently have people with this title, *this is a new competency*.

While companies often have people with the title "product manager," that title can be very misleading as true product teams have very different demands on the product manager than the feature teams they are replacing.

Here is where experienced product leaders are so essential. These leaders need to determine which people are more suited to other roles and which have the potential through coaching and training to become the true product managers that the stakeholders need to truly partner with.

To be explicit, the senior leadership team should believe that each product manager is a potential future leader of the company.

To be even more explicit, product leaders will be judged based on their weakest product manager.

This product leadership is not a minor role, and it requires strong people with a deep understanding of your customers, the data, your business, the market, and the technology.

5. Professional Product Designers

As the product model puts users and customers much closer to the product teams and to the company, product designers have the skills to help you craft customer experiences that your customers love. This is why we elevate design from a supporting role to a central role, alongside product managers and engineers.

6. Empowered Engineers

Empowered engineers form the engine that fuels consistent innovation. Much of effective transformation is all about enabling and encouraging truly empowered engineers.

Just in case it's not blatantly obvious, you won't have empowered engineers if you outsource your engineers.

No one is suggesting that you put your engineers on a pedestal, but you need to get them out of the proverbial basement and put them front and center on your product teams and in coming up with solutions to the hardest problems you face.

7. Insights-Based Product Strategy

The purpose of your product strategy is to determine the most important problems that need to be solved for the company to achieve its objectives.

Most companies that have yet to transform have never created a product strategy before. That's because their strategy was simply to serve as many of the business stakeholders as they could. And, of course, that's not a strategy at all.

Yet an effective product strategy, based on both quantitative and qualitative insights, is key not only to leveraging the talents of your people but to getting the most out of your technology investment.

Your product leaders are responsible for this product strategy—yet another reason why it's so important to have experienced and strong product leaders.

8. Stakeholder Collaboration

One of the most difficult aspects of a successful transformation is redefining the relationship between the product organization and the different parts of the business.

The reason it's difficult is because it represents a real change for many key stakeholders.

Fundamentally, transformation involves moving from a model where the technology teams exist to serve the business to one where they exist to serve the customers, in ways that work for the business.

To be explicit, that means moving from a *subservient* model to a *collaborative* model.

Most stakeholders are frustrated with the old way of working, so they're at least willing to try to transform, but some likely will take the loss of control personally. Product leaders and product teams need to be sensitive to this issue.

It's essential that product leaders do not try to push this change until and unless product teams have competent product managers who are ready to step up and do what they need to. But once these competent product managers are in place, they may need the active support of senior leadership to help stakeholders through this change to the collaborative model.

It's important not to sugarcoat this, as the change is profound.

9. Continuous Evangelization of Outcomes

Yet another critical responsibility of your product leaders is continuous evangelism. They need to evangelize the product vision, the product strategy, the importance of moving the focus to outcomes, and the broader transformation to the product model.

The product vision is the inspiration and motivation for the work and combined efforts of the whole organization.

The product strategy is critically important to evangelize because it makes transparent the rationale and the evidence for what will get done and what won't.

More generally, product teams are learning every week as they work on product discovery for the problems they are being asked to solve.

It's important to share the learnings honestly and openly so that the broader company can understand how learning happens, where the innovations are coming from, and what it means to truly deliver business results versus just shipping features and projects.

10. Corporate Courage

There is no question that successful transformation is difficult. We hope that this list helps illuminate specifically why that's the case and what needs to be tackled in order to succeed.

However, the list would not be complete without acknowledging that every case of successful transformation we know of required true courage from executives and other senior leaders.

Moving to a fundamentally different model—even when the current model is broken—requires a real leap of faith, and it takes courage to make the leap. Often, the very senior leaders of companies that transform don't get the recognition for this courage they deserve (although the stock market has rewarded them very well). But still, we know many more senior leaders who don't have the courage and aren't willing to do what it takes to help their company not just survive, but thrive.

We hope that these ten keys provide a deeper sense of what's required to successfully and effectively transform your company to the product model.

But What Can I Do?

We realize that, in many cases, the first person in a company to read this book may be an individual contributor or one of the product leaders.

It may be very clear to you that your company is in desperate need of change, yet you might feel like you're just shouting into the wind.

It is true that to succeed in moving to the product model, you really are going to need the active help and support of the leaders of the company, or at least your business unit.

While we hope this book finds its way into the hands of many more CEOs, there are still things that you can do, even as an individual contributor.

Here's what we suggest:

1. Talk to the rest of your product team to see if they are up for doing this with you. If it's just you, then your first job is to try to convince your team of the benefits.

2. Once they are with you, focus on upskilling yourselves to the new product model competencies described in the book. At an absolute minimum, learning these new skills will only help your careers.

3. Once you feel like you have some new skills and you'd like to try them out, go to your manager and ask if the company would be willing to let your product team serve as a pilot for this new way of working for the next quarter or two.

4. Suggest you focus on changing how you build and changing how you solve problems.

5. If the experiment goes well, hopefully it will spread. If it doesn't, you have contained the risk to just your team.

CHAPTER

49

Innovation Story: Trainline

Marty's Note: As you saw earlier, the Trainline product leadership built a very strong product organization, capable of the consistent product innovation they needed. Of all the examples of their skills, this story is my favorite. You can see each of the product model competencies and product model concepts in action as they tackle a very difficult problem, but one that promised real value for their customers, their rail partners, and their company. The level of product innovation described here rivals anything I have seen at the world's most famous tech product companies.

Company Background

After working through the transformation described in part V, Trainline now had capable engineering, product management, and design, along with a strong product vision and product strategy. The company also was starting to see real payoff from its investment in data science.

Trainline had built successful and effective real-time train location, estimated platform departures, personalized disruption alerts, and the ability to find less congested carriages, all of which helped to bring their vision to life.

But one major issue in particular continued to frustrate and concern their customers: pricing.

The Problem to Solve

From an early stage, it was clear that the perceived expense of rail tickets was hindering Trainline's ability to attract more customers. Customers would tell them, over and over, "Tickets are so expensive!"

Conversely, the industry would counter that this was simply not true, pointing to the fact that tickets for every route were on sale at extremely low prices. Understanding this apparent schism had the potential to result in further strong innovation, but with millions of potential journey possibilities, it would also be significant work.

It was also an issue that, on the face of it, the company had little control over: Trainline received prices in real time the moment a customer executed a search. Only then would an operator's API expose the price for that specific individual journey.

However, Trainline's head of data science was convinced that such an important problem was worth an investigation.

Discovering the Solution

What the team discovered gave another truly unique insight: Yes, tickets were cheap, but not so cheap that a significant percentage of customers wanted, or needed, to buy them.

Initially starting with a smaller data set and then validating the conclusions at scale, the team proved that prices were *not* linked to yield. They were instead linked directly to days to departure. Buying early meant cheaper tickets were available, but purchasing later—within two weeks of a planned departure—resulted in steep rises. The increases uncovered sometimes were eye watering.

While the industry was loudly promoting the lowest-priced tickets, the vast majority of customers had to pay significantly more.

The team could find no evidence of smart yield management or automated last-minute sell-offs. This was a significant insight—and potentially a big win. The team was now able to understand the exact price savings for buying tickets early, on any route.

The question now was: Could they figure out a way of showcasing this information to customers in real time, helping them make more informed choices and saving money?

While the opportunity was clear, the viability risks were significant. Trainline was a large and growing part of a much wider industry. Its rail partners were critical to its mission. Shouting loudly about Trainline's newfound ability had maximum potential to alienate those the company relied on to trade.

Would the solution be valuable and usable enough to result in a real-world change of behavior? Would Trainline be able to scale it accurately? With so much information baked into the mobile real estate, could the company design an experience that made sense?

But above all, the significant question remained whether Trainline would be able to launch without causing its partners—the rail companies—to convulse.

This had maximum potential to blow up. Showcasing erroneous ticket prices could result in the immediate withdrawal of the company's license to trade. For that reason alone, the executive team—including the CEO—was brought into the viability conversation during discovery.

The data scientists responsible for data discovery now joined forces with the app team, moving into a comprehensive product discovery effort.

Addressing value risk first, the team hard-coded a thinly sliced solution, testing it in real time on a few select routes. The live-data prototype achieved unanimously positive feedback. These were real savings, and customers loved it. The team's confidence grew.

In line with the desire for a premium mobile experience, the team had decided to make this solution unique to Trainline's apps. This directly reflected the company's vision and strongly supported its product strategy.

Thanks to a highly capable design team, and intensive iteration and user prototyping, usability was unlikely to pose a major threat. But the bigger risks remained: feasibility and viability.

Accuracy was critical. This was a complex, big-data problem, even for a team with significant smarts. The team set the minimum bar in the high nines—there could be no risk taken with losing Trainline's license. The team worked diligently, and their ability to execute with extreme accuracy increased, bolstered by their rapid ability to performantly scale on AWS.

One significant risk remained: viability. While Trainline did not formally require approval from its partners, neither the team, the executives, nor the board would contemplate severe blowback from the industry, especially so close to a potential IPO.

At the very least, exposing the pricing strategy that sat behind UK rail tickets would be controversial. Trainline could prove that it would result in customer savings, but given the company's now-significant market share, leaders started to worry that their solution could result in net industry revenue decline.

They required real data, at scale. They were going to need to lean heavily on their small but expert operational team, comprising individuals who had spent years developing close relationships with their industry partners. Without them, the initiative likely was doomed to failure.

Their ops team members were excited, but nervous, suggesting they focus their immediate attention on a smaller number of partners with a more progressive digital outlook. This work was nuanced and complex. Many partners refused, frustrated at yet another intervention from the upstart of the industry. But one agreed to a live test. Finally, the team had an opportunity to quietly launch and watch how the real data played out.

The Results

The conclusions were extremely positive. As expected, the cohort of customers with increased flexibility started to purchase tickets earlier, making real savings and increasing their usage. They worked through the data carefully, confirming that the additional revenue derived from filling more seats more than offset the lower prices.

Those who had saved money were ecstatic. The test had been an overwhelming success, allowing team members to make a firmer case for a (still carefully managed) wider rollout, which generated another chorus of positive free publicity.

The release spoke directly to Trainline's mission: to promote lower carbon transport solutions by offering the cheapest possible prices. It also gave the marketing team a significant new opportunity to showcase Trainline as the sole innovator within the sector. The various campaigns run were both timely, given Timeline's impending public listing, and highly effective. National media picked it up immediately, and a growing number of technology sites regularly included Trainline in their daily coverage.

Internally, this success also gave the company another strong shot of confidence. Its price-prediction tool was just the latest example of a growing list of solutions that reflected Trainline's strong and maturing technology culture.

Release by release, the company had achieved significant incremental value—not just for customers and the business but for shareholders. Just in time for the forthcoming IPO.

Learning More

One of the limitations of a physical book is that once it's printed, it's not something we can update. Yet we know we will continue to receive new questions and encounter new objections. If you have a question that is not discussed in this book, please see our living repository of questions pertaining to this book (see https://svpg.com/transformed-faq).

The Silicon Valley Product Group website (https://svpg.com/) is designed as a free and open resource where we share our latest thoughts, learnings, and examples from companies operating in the product model.

SVPG also holds workshops—both online and in-person—for product managers, product teams, product leaders, and company executives who wish to learn how to operate in the product model (see https://svpg.com/workshops/).

For companies that believe they need meaningful transformation across their technology and product organization to competitively produce technology-powered products, we also offer custom, on-site engagements.

Acknowledgments

This book is based on the lessons learned from the SVPG partners as we've spent the past 20 years helping companies move to the product model.

We wanted to share what we've learned far beyond the relative handful of companies that we are able to work with personally.

As we said earlier, nothing in this book was invented by us. We simply share what we have seen work in the top product companies, and we help others adopt those principles and behaviors.

Much of our content is informed by the global product community. When we publish articles, give conference talks, and host webinars and workshops, we test out many of the techniques and case studies that we write about. People have always been generous with their feedback, and many send us additional questions. Much of this book was inspired by these interactions.

While much of what we write about is inspired by the community, we work very hard to make sure we explain the concepts in the best way we can, and for that we depend on a set of expert reviewers. A very big thank you to Shawn Boyer, Matt Brown, Gabi Bufrem, Felipe Castro, Shreyas Doshi, Mike Fisher, Chuck Geiger, Stacey Langer, Michele Longmire, and Alex Pressland. Each and every one of them left their mark on this book.

We would also like to thank the product coaches who allowed us to profile them in this book: Gabi Bufrem, Hope Gurion, Margaret Hollendoner, Stacey Langer, Marily Nika, Phyl Terry, and Petra Wille.

And on a personal note, this book would not be possible without my SVPG partners: Lea Hickman, Christian Idiodi, Chris Jones, Martina Lauchengco, and Jon Moore. Each of them contributed original content and countless suggestions to this book. I must also specifically thank Chris, as he has literally been with me every single step of the way, and has become an integral and indispensable part of my writing process. Thanks also to my longtime editor, Peter Economy, and the publishing team at John Wiley & Sons. And finally, to Lynn for her love and support through now four major writing projects.

About the Authors

This book has been written by the five SVPG product partners.

Our belief is that to lead an effective transformation to the product model, it is critical to have personally been there and done that, and to truly know what good looks like.

That is why each partner has built products for decades, first as a product creator, and then as a product leader, at many of the most successful tech-powered product companies in the world.

As a product partner for SVPG, we work with product organizations of all sizes, stages, and industries in order to help them learn how to leverage technology to create effective solutions for their customers.

We do not employ junior partners or intermediaries. We also do not do the work for you. We work personally and directly with the people at every level of a company to help them learn the necessary skills so that they can lead their companies into the future.

You can learn more about the individual partners at www.svpg .com/team.